Practical Apache Spark

Using the Scala API

Subhashini Chellappan

Dharanitharan Ganesan

Apress®

Practical Apache Spark

Subhashini Chellappan
Bangalore, India

Dharanitharan Ganesan
Krishnagiri, Tamil Nadu, India

ISBN-13 (pbk): 978-1-4842-3651-2   ISBN-13 (electronic): 978-1-4842-3652-9
https://doi.org/10.1007/978-1-4842-3652-9

Library of Congress Control Number: 2018965197

Managing Director, Apress Media LLC: Welmoed Spahr
Acquisitions Editor: Celestin Suresh John
Development Editor: Siddhi Chavans
Coordinating Editor: Aditee Mirashi

Cover image by Freepik (www.freepik.com)

Distributed to the book trade worldwide by Springer Science+Business Media New York, 233 Spring Street, 6th Floor, New York, NY 10013. Phone 1-800-SPRINGER, fax (201) 348-4505, e-mail orders-ny@springer-sbm.com, or visit www.springeronline.com. Apress Media, LLC is a California LLC and the sole member (owner) is Springer Science + Business Media Finance Inc (SSBM Finance Inc). SSBM Finance Inc is a **Delaware** corporation.

For information on translations, please e-mail rights@apress.com, or visit http://www.apress.com/rights-permissions.

Apress titles may be purchased in bulk for academic, corporate, or promotional use. eBook versions and licenses are also available for most titles. For more information, reference our Print and eBook Bulk Sales web page at http://www.apress.com/bulk-sales.

Any source code or other supplementary material referenced by the author in this book is available to readers on GitHub via the book's product page, located at www.apress.com/978-1-4842-3651-2. For more detailed information, please visit http://www.apress.com/source-code.

Printed on acid-free paper

Table of Contents

About the Authors ix

About the Technical Reviewers xi

Acknowledgments xiii

Introduction xv

Chapter 1: Scala: Functional Programming Aspects 1

What Is Functional Programming? 2

    What Is a Pure Function? 2

    Example of Pure Function 3

Scala Programming Features 4

    Variable Declaration and Initialization 5

    Type Inference 6

    Immutability 7

    Lazy Evaluation 8

    String Interpolation 10

    Pattern Matching 13

    Scala Class vs. Object 14

    Singleton Object 15

    Companion Classes and Objects 17

    Case Classes 18

    Scala Collections 21

Functional Programming Aspects of Scala 27

    Anonymous Functions 27

    Higher Order Functions 29

    Function Composition 30

    Function Currying 31

Nested Functions 32

Functions with Variable Length Parameters 34

Reference Links 37

Points to Remember 37

Chapter 2: Single and Multinode Cluster Setup 39

Spark Multinode Cluster Setup 39

Recommended Platform 39

Prerequisites 61

Spark Installation Steps 62

Spark Web UI 66

Stopping the Spark Cluster 70

Spark Single-Node Cluster Setup 70

Prerequisites 71

Spark Installation Steps 73

Spark Master UI 76

Points to Remember 77

Chapter 3: Introduction to Apache Spark and Spark Core 79

What Is Apache Spark? 80

Why Apache Spark? 80

Spark vs. Hadoop MapReduce 81

Apache Spark Architecture 82

Spark Components 84

Spark Core (RDD) 84

Spark SQL 84

Spark Streaming 85

MLib 85

GraphX 85

SparkR 85

Spark Shell 85

Spark Core: RDD 86

RDD Operations 88

Creating an RDD 88

RDD Transformations 91

RDD Actions 95

Working with Pair RDDs 98

Direct Acylic Graph in Apache Spark.. 101

How DAG Works in Spark.. 101

How Spark Achieves Fault Tolerance Through DAG... 103

Persisting RDD 104

Shared Variables 105

Broadcast Variables.. 106

Accumulators 106

Simple Build Tool (SBT) 107

Assignments 112

Reference Links 112

Points to Remember 113

Chapter 4: Spark SQL, DataFrames, and Datasets 115

What Is Spark SQL? 116

Datasets and DataFrames 116

Spark Session 116

Creating DataFrames 117

DataFrame Operations.. 118

Running SQL Queries Programatically.. 121

Dataset Operations 123

Interoperating with RDDs 125

Different Data Sources 129

Working with Hive Tables 133

Building Spark SQL Application with SBT.. 135

Points to Remember 139

Chapter 5: Introduction to Spark Streaming 141

Data Processing 142

Streaming Data 142

    Why Streaming Data Are Important 142

Introduction to Spark Streaming 142

    Internal Working of Spark Streaming 143

    Spark Streaming Concepts 144

Spark Streaming Example Using TCP Socket 145

Stateful Streaming 149

    Window-Based Streaming 149

    Full-Session-Based Streaming 152

Streaming Applications Considerations 155

Points to Remember 156

Chapter 6: Spark Structured Streaming 157

What Is Spark Structured Streaming? 158

Spark Structured Streaming Programming Model 158

    Word Count Example Using Structured Streaming 160

Creating Streaming DataFrames and Streaming Datasets 163

Operations on Streaming DataFrames/Datasets 164

Stateful Streaming: Window Operations on Event-Time 167

Stateful Streaming: Handling Late Data and Watermarking 170

Triggers 171

Fault Tolerance 173

Points to Remember 174

Chapter 7: Spark Streaming with Kafka 175

Introduction to Kafka 175

    Kafka Core Concepts 176

    Kafka APIs 176

Kafka Fundamental Concepts 177

Kafka Architecture 178

Kafka Topics 179

Leaders and Replicas 179

Setting Up the Kafka Cluster 180

Spark Streaming and Kafka Integration 182

Spark Structure Streaming and Kafka Integration 185

Points to Remember 187

Chapter 8: Spark Machine Learning Library 189

What Is Spark MLlib? 190

Spark MLlib APIs 190

Vectors in Scala 191

Basic Statistics 194

Extracting, Transforming, and Selecting Features 200

ML Pipelines 215

Points to Remember 236

Chapter 9: Working with SparkR 237

Introduction to SparkR 237

SparkDataFrame 237

SparkSession 238

Starting SparkR from RStudio 238

Creating SparkDataFrames 241

From a Local R DataFrame 241

From Other Data Sources 242

From Hive Tables 243

SparkDataFrame Operations 244

Selecting Rows and Columns 244

Grouping and Aggregation 245

Operating on Columns 247

Applying User-Defined Functions 248

Run a Given Function on a Large Data Set Using dapply or dapplyCollect 248

Running SQL Queries from SparkR 249

Machine Learning Algorithms 250

    Regression and Classification Algorithms 250

    Logistic Regression 255

    Decision Tree 258

Points to Remember 260

Chapter 10: Spark Real-Time Use Case **261**

Data Analytics Project Architecture 262

    Data Ingestion 262

    Data Storage 263

    Data Processing 263

    Data Visualization 264

Use Cases 264

    Event Detection Use Case 264

    Build Procedure 270

    Building the Application with SBT 271

Points to Remember 273

Index **275**

About the Authors

Subhashini Chellappan is a technology enthusiast with expertise in the big data and cloud space. She has rich experience in both academia and the software industry. Her areas of interest and expertise are centered on business intelligence, big data analytics and cloud computing.

Dharanitharan Ganesan has an MBA in technology management with a high level of exposure and experience in big data, using Apache Hadoop, Apache Spark, and various Hadoop ecosystem components. He has a proven track record of improving efficiency and productivity through the automation of various routine and administrative functions in business intelligence and big data technologies. His areas of interest and expertise are centered on machine learning algorithms, Blockchain in big data, statistical modeling, and predictive analytics.

About the Technical Reviewers

Mukund Kumar Mishra is a senior technologist with strong business acumen. He has more than 18 years of international experience in business intelligence, big data, data science, and computational analytics. He is a regular speaker on big data concepts, Hive, Hadoop, and Spark. Before joining the world of big data, he also worked extensively in the Java and .NET space.

Mukund is also a poet and his first book of poetry was published when he was only 15 years old. Thus far he has written around 300 poems. He runs one of the largest Facebook groups on big data Hadoop (see `https://www.facebook.com/groups/656180851123299/`). You can connect with Mukund on LinkedIn at `https://www.linkedin.com/in/mukund-kumar-mishra-7804b38/`.

Sundar Rajan Raman has more than 14 years of full stack IT experience, including special interests in machine learning, deep learning, and natural language processing. He has 6 years of big data development and architecture experience including Hadoop and its ecosystems and other No SQL technologies such as MongoDB and Cassandra. He is a design thinking practitioner interested in strategizing using design thinking principles.

Sundar is active in coaching and mentoring people. He has mentored many teammates who are now in respectable positions in their careers.

Acknowledgments

The making of this book was a journey that we are glad we undertook. The journey spanned a few months, but the experience will last a lifetime. We had our families, friends, collegues, and well-wishers onboard for this journey, and we wish to express our deepest gratitude to each one of them.

We would like to express our special thanks to our families, friends, and colleagues, who provided that support that allowed us to complete this book within a limited time frame.

Special thanks are extended to our technical reviewers for the vigilant review and filling in with their expert opinion.

We would like to thank Celestin Suresh John, Senior Manager, Apress and Springer Science and Business Media, for signing us up for this wonderful creation. We wish to acknowledge and appreciate Aditee Mirashi, coordinating editor, and the team who guided us through the entire process of preparation and publication.

Introduction

Why This Book?

Apache Spark is a fast, open source, general-purpose memory processing engine for big data processing. This book discusses various components of Apache Spark, such as Spark Core, Spark SQL DataFrames and Datasets, Spark Streaming, Structured Streaming, Spark machine learning libraries, and SparkR with practical code snippets for each module. It also covers the integration of Apache Spark with other ecosystem components such as Hive and Kafka. The book has within its scope the following:

* Functional programming features of Scala.

* Architecture and working of different Spark components.

* Work on Spark integration with Hive and Kafka.

* Using Spark SQL DataFrames and Datasets to process the data using traditional SQL queries.

* Work with different machine learning libraries in Spark MLlib packages.

Who Is This Book For?

The audience for this book includes all levels of IT professionals.

How Is This Book Organized?

Chapter 1 describes the functional programming aspects of Scala with code snippets. In Chapter 2, we explain the steps for Spark installation and cluster setup. Chapter 3 describes the need for Apache Spark and core components of Apache Spark. In Chapter 4, we explain how to process structure data using Spark SQL, DataFrames, and Datasets. Chapter 5 provides the basic concepts of Spark Streaming and Chapter 6 covers the basic concepts of Spark Structure Streaming. In Chapter 7, we describe how to integrate Apache Spark with Apache Kafka. Chapter 8 then explains the machine learning library of Apache Spark. In Chapter 9, we address how to integrate Spark with R. Finally, in Chapter 10 we provide some real-time use cases for Apache Spark.

How Can I Get the Most Out of This Book?

It is easy to leverage this book for maximum gain by reading the chapters thoroughly. Get hands-on by following the step-by-step instructions provided in the demonstrations. Do not skip any of the demonstrations. If need be, repeat them a second time or until the concept is firmly etched in your mind. Happy learning!!!

Subhashini Chellappan
Dharanitharan Ganesan

CHAPTER 1

Scala: Functional Programming Aspects

This chapter is a prerequiste chapter that provides a high-level overview of functional programming aspects of Scala. This chapter helps you understand the functional programming aspects of Scala. Scala is a preferred language to work with Apache Spark. After this chapter, you will be able to understand the building blocks of functional programming and how to apply functional programming concepts in your daily programming tasks. There is a hands-on focus in this chapter and the entire chapter introduces short programs and code snippets as illustrations of important functional programming features.

The recommended background for this chapter is some prior experience with Java or Scala. Experience with any other programming language is also sufficient. Also, having some familiarity with the command line is preferred.

By end of this chapter, you will be able to do the following:

- Understand the essentials of functional programming.
- Combine functional programming with objects and classes.
- Understand the functional programming features.
- Write functional programs for any programming tasks.

Note   It is recommended that you practice the code snippets provided and practice the exercises to develop effective knowledge of the functional programming aspects of Scala.

© Subhashini Chellappan, Dharanitharan Ganesan 2018
S. Chellappan and D. Ganesan, *Practical Apache Spark*, https://doi.org/10.1007/978-1-4842-3652-9_1

What Is Functional Programming?

Functional programming (FP) is a way of writing computer programs as the evaluation of mathematical functions, which avoids changing the state or mutating data. The programs are constructed using pure functions. Functional programs are always declarative, where the programming is done with declarations and expressions instead of statements. Functional programming languages are categorized into two groups:

1. Pure function

2. Impure function

What Is a Pure Function?

A function that has no side effects is called a pure function. So, what are side effects? A function is said to be having side effects if it does any of the following other than just returning a result:

- Modifies an existing variable.

- Reads from a file or writes to a file.

- Modifies a data structure (e.g., array, list).

- Modifies an object (setting a field in an object).

The output of a pure function depends only on the input parameter passed to the function. The pure function will always give the same output for the same input arguments, irrespective of the number of times it is called.

The impure function can give different output every time it is called and the output of the function is not dependent only on the input parameters.

Hint   Let us try to understand pure and impure functions using some Java concepts (if you are familiar with). The mutator method (i.e., the setter method) is an impure function and the accessor method (i.e., the getter method) is a pure function.

Example of Pure Function

The following function is an example of a pure function:

```
def squareTheNumber(num : Int) :Int ={
    return num*num
}
```

The function squareTheNumber (see Figure 1-1) accepts an integer parameter and always returns the square of the number. Because it has no side effects and the output is dependent only on the input parameter, it is considered a pure function.

```
scala> def squareTheNumber(num : Int) :Int ={
     | return num*num
     | }
squareTheNumber: (num: Int)Int

scala> squareTheNumber(10)
res3: Int = 100

scala>
```

Figure 1-1. *Example of a pure function*

Here are some the typical examples of pure functions:

- Mathematical functions such as addition, subtraction, division, and multiplication.

- String class methods like length, toUpper, and toLower.

These are some typical examples of impure functions:

- A function that generates a random number.

- Date methods like getDate() and getTime() as they return different values based on the time they are called.

PURE AND IMPURE FUNCTIONS EXERCISE

---

1. Find the type of function and give the reason.

```scala
def myFunction(a : Int) :Int ={
   return a
}
```

2. Find the type of function and give the reason.

```scala
def myFunction() : Double = {
   var a = Math.random()
   return a
}
```

3. The following function is said to be an impure function. Why?

```scala
def myFunction(emp : Employee) : Double = {
   emp.setSalary(100000)
   return emp.getSalary()
}
```

4. Give five differences between pure functions and impure functions.

5. A function named `acceptUserInput()` contains a statement to get input from the console. Identify whether the function is pure or impure and justify the reason.

Note   The last statement of the function is always a `return` statement in Scala. Hence, it is not necessary to explicitly specify the `return` keyword.

The semicolon is not needed to specify the end of a statement in Scala. By default, the newline character (\n) is considered the end of a statement. However, a semicolon is needed if multiple statements are to be written in a single line.

Scala Programming Features

Let us turn to the Scala programming features, as illustrated in 1-2.

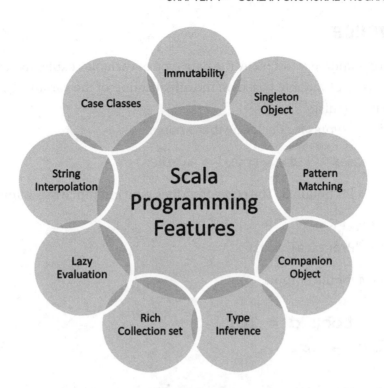

Figure 1-2. *Features of Scala programming language*

Variable Declaration and Initialization

The variables can be declared through var and val keywords. The difference between var and val is explained later in this chapter. The code here describes val and var:

```
val bookId=100
var bookId=100
```

Figure 1-3 displays the output.

```
scala> val bookId = 100
bookId: Int = 100


scala> var bookId = 100
bookId: Int = 100
```

Figure 1-3. *Variable declaration and initialization*

Type Inference

In Scala, it is not mandatory to specify the data type of variables explicitly. The compiler can identify the type of variable based on the initialization of the variable by the built-in type inference mechanism.

The following is the syntax for declaring the variable:

```
var <variable_name> : [<data_type>] = <value>
```

The [<data_type>] is optional. The code describes type inference mechanism.

```
var bookId = 101
var bookName = "Practical Spark"
```

Refer to Figure 1-4 for the output.

```
scala> var bookId = 101
bookId: Int = 101

scala> var bookName = "Practical Spark"
bookName: String = Practical Spark
```

Figure 1-4. *Type inference without an explicit type specification*

However, you can explicity specify the type for variables during declaration as shown here:

```
var bookId:Int = 101
var bookName:String = "Practical Spark"
```

Figure 1-5 shows the output.

```
scala> var bookId:Int = 101
bookId: Int = 101

scala> var bookName:String = "Practical Spark"
bookName: String = Practical Spark
```

Figure 1-5. *Type inference with an explicit type specification*

Immutability

Immutablity means the value of a variable cannot be changed once it is declared. The keyword val is used to declare immutable variables, whereas mutable variables can be declared using the keyword var. Data immutablity helps you achieve concurrency control while managing data. The following code illustrates a mutable variable.

```
var bookName = "Spark"
bookName = "Practical Spark"
print("The book Name is" + bookName)
```

Figure 1-6 shows mutable variables.

```
scala> var bookName = "Spark"
bookName: String = Spark

scala> bookName = "Practical Spark"
bookName: String = Practical Spark

scala> print("The book Name is " + bookName)
The book Name is Practical Spark
```

Figure 1-6. *Mutable variables using the* var *keyword*

Hence, variable reassignment is possible if the variable is declared using the var keyword. The code shown here illustrates an immutable variable.

```
val bookName = "Spark"
bookName = "Practical Spark"
```

Refer to Figure 1-7 for immutable variables.

```
scala> val bookName = "Spark"
bookName: String = Spark

scala> bookName = "Practical Spark"
<console>:15: error: reassignment to val
        bookName = "Practical Spark"
                 ^
```

Figure 1-7. *Immutable variables using the* val *keyword*

As you can see, variable reassignment is not possible if the variable is declared using the val keyword.

Hint   Declaring immutable variables using the val keyword is like declaring final variables in Java.

Lazy Evaluation

The lazy evaluation feature allows the user to defer the execution of any expression until it is needed using the lazy keyword. When the expression is declared with the lazy keyword, it will be executed only when it is being called explicity. The following code and Figure 1-8 illustrates immediate expression evaluation.

```
val x = 10
val y = 10
val sum = x+y
```

```
scala> val x = 10
x: Int = 10

scala> val y = 10
y: Int = 10

scala> val sum = x+y
sum: Int = 20
```

Figure 1-8. *Immediete expression evaluation without the* `lazy` *keyword*

In the following code the expression y is defined with the `lazy` keyword. Hence, it is evaluated only when it is called. Refer to Figure 1-9 for the output.

```
val x = 10
val y = 10
lazy val y = 10
print(sum)

scala> val x = 10
x: Int = 10

scala> val y = 10
y: Int = 10

scala> lazy val y = 10
y: Int = <lazy>

scala> print (sum)
20
```

Figure 1-9. *Lazy evaluation with the* `lazy` *keyword*

It is important to note that the lazy evaluation feature can be used only with val (i.e., immutable variables). Refer to the code given here and Figure 1-10.

```
var x =10
var y =10
lazy sum = x+y
```

```
scala> var x = 10
x: Int = 10

scala> var y = 10
y: Int = 10

scala> lazy sum = x+y
<console>:1: error: lazy not allowed here. Only vals can be lazy
        lazy sum = x+y
             ^
```

Figure 1-10. *Lazy evaluation cannot be used with mutable variables*

String Interpolation

String interpolation is the process of creating a string from the data. The user can embed the references of any variable directly into the processed string literals and format the string. The code shown here describes string processing without using string interpolation.

```
var bookName = "practical Spark"
println("The Book name is" + bookName)
```

Refer to Figure 1-11 for the output

```
scala> var bookName = "Practical Spark"
bookName: String = Practical Spark

scala> println("The Book name is "+bookName)
The Book name is Practical Spark
```

Figure 1-11. *String processing without using interpolation*

These are the available string interpolation methods:

- s interpolator.

- f interpolator.

- raw interpolator.

String - s Interpolator

Using the interpolator s, to the string literal allows the user to use the reference variables to append the data directly. The following code illustrates the s interpolator and the result is shown in Figure 1-12.

```
var bookName = "practical Spark"
println(s"The Book name is $bookName")
```

```
scala> var bookName = "Practical Spark"
bookName: String = Practical Spark

scala> println(s"The Book name is $bookName")
The Book name is Practical Spark
```

Figure 1-12. *String processing using the s interpolator*

Observe the difference in println method syntax to form the string with and without string interpolation.

Also, the arbitary expressions can be evaluated using the string interpolators, as shown in the following code. Refer to Figure 1-13 for the output.

```
val x = 10
val y =15
println(s"The sum of $x and $y is ${x+y}")
```

```
scala> val x = 10
x: Int = 10

scala> val y = 15
y: Int = 15

scala> println(s"The sum of $x and $y is ${x+y}")
The sum of 10 and 15 is 25
```

Figure 1-13. *Expression evaluation using string interpolation*

String - f Interpolator

Scala offers a new mechanism to create strings from your data. Using the interpolator f to the string literal allows the user to create the formatted string and embed variable references directly in *processed* string literals. The following code illustrates the f interpolator and the output is shown in Figure 1-14.

```
var bookPrice = 100
val bookName = "Practical Spark"
println(f"The price of $bookName is $bookPrice")
println(f"The price of $bookName is $bookPrice%1.1f")
println(f"The price of $bookName is $bookPrice%1.2f")
```

```
scala> val bookPrice = 100
bookPrice: Int = 100

scala> val bookName = "Practical Spark"
bookName: String = Practical Spark

scala> println(f"The price of $bookName is $bookPrice")
The price of Practical Spark is 100

scala> println(f"The price of $bookName is $bookPrice%1.1f")
The price of Practical Spark is 100.0

scala> println(f"The price of $bookName is $bookPrice%1.2f")
The price of Practical Spark is 100.00
```

Figure 1-14. *String processing using the f interpolator*

The formats allowed after % are based on string format utilities available from Java.

String - raw Interpolator

The raw interpolator does not allow the escaping of literals. For example, using \n with the raw interpolator does not return a newline character. The following code illustrates the raw interpolator and the output is shown in Figure 1-15.

```
val bookId = 101
val bookName = "Practical Spark"
println(s"The book id is $bookId. \n The book name is $bookName")
println(raw"The id is $bookId. \n The book name is $bookName")
```

```
scala> val bookId = 101
bookId: Int = 101

scala> val bookName = "Practical Spark"
bookName: String = Practical Spark

scala> println(s"The book id is $bookId. \n The book name is $bookName")
The book id is 101.
 The book name is Practical Spark

scala> println(raw"The book id is $bookId. \n The book name is $bookName")
The book id is 101. \n The book name is Practical Spark
```

Figure 1-15. *String processing using the* raw *interpolator*

Pattern Matching

The process of checking a pattern against a value is called pattern matching. A successful match returns a value associated with the case. Here is the simple syntax to use pattern matching.

```
<reference_name> match {
  case <option 1> => <return_value 1>
  case <option 2> => <return_value 2>
  case <option n> => <return_value n>
  case <default_option> => <default return_value>
}
```

The pattern matching expression can be defined for a function as shown here.

```
def chapterName(chapterNo:Int) = chapterNo match {
    case 1 => "Scala Features"
    case 2 => "Spark core"
    case 3 => "Spark Streaming"
    case _ => "Chapter not defined"
    }
```

Refer to Figure 1-16 for the output.

```
scala> def chapterName(chapterNo:Int) = chapterNo match {
     | case 1 => "Scala Features"
     | case 2 => "Spark core"
     | case 3 => "Spark Streaming"
     | case _ => "Chapter not defined"
     | }
chapterName: (chapterNo: Int)String

scala> chapterName(1)
res30: String = Scala Features

scala> chapterName(5)
res31: String = Chapter not defined
```

Figure 1-16. *Example for pattern matching*

Scala Class vs. Object

A class is a collection of variables, functions, and objects that is defined as a blueprint for creating objects (i.e., instances). A Scala class can be instantiated (object can be created). The following code describes class and objects.

```
scala> class SparkBook {
     |         val bookId = 101
     |         val bookName = "Practical Spark"
     |         val bookAuthor = "Dharanitharan G"
     |         def printBookDetails(){
     |             println(s"The $bookName is written by $bookAuthor")
     | }
     | }
```

```
defined class SparkBook

scala> val book = new SparkBook()
book: SparkBook = SparkBook@96be74

scala> book.printBookDetails()
The Practical Spark is written by Dharanitharan G
```

Figure 1-17 displays the output.

```
scala> class SparkBook {
       | val bookId = 101
       | val bookName = "Practical Spark"
       | val bookAuthor = "Dharanitharan G"
       | def printBookDetails(){
       |   println(s"The $bookName is written by $bookAuthor")
       | }
       | }
defined class SparkBook

scala> val book = new SparkBook()
book: SparkBook = SparkBook@762af0e9

scala> book.printBookDetails()
The Practical Spark is written by Dharanitharan G
```

Figure 1-17. *Example for class and objects*

The functions in the class can be called by using the object reference. The new keyword is used to create an object, or instance of the class.

Singleton Object

Scala classes cannot have static variables and methods. Instead, a Scala class can have a singleton object or companion object. You can use singleton object when there is a need for only one instance of a class. A singleton is also a Scala class but it has only one instance (i.e., Object). The singleton object cannot be instantiated (object creation). It can be created using the object keyword. The functions and variables in the singleton object can be directly called without object creation. The code shown here describes SingletonObject and the output is displayed in Figure 1-18.

```
scala> object SingletonObjectDeo{
     | def functionInSingleton{
     | println("This is printed in Singleton Object")
     | }
     | }
```

```
scala> object SingletonObjectDemo {
       | def functionInSingleton() {
       | println("This is printed in Singleton Object")
       | }
       | }
defined object SingletonObjectDemo

scala> SingletonObjectDemo.functionInSingleton()
This is printed in Singleton Object
```

Figure 1-18. *Singleton object*

Generally, the `main` method is created in a singleton object. Hence, the compiler need not create an object to call the `main` method while executing. Add the following code in a `.scala` file and execute it in a command prompt (REPL) to understand how the Scala compiler calls the `main` method in singleton object.

```
object SingletonObjectMainDemo {
    def main(args: Array[String]) {
    println("This is printed in main method")
    }
}
```

Save this code as `SingletonObjectMainDemo.scala` and execute the program using these commands at the command prompt.

```
scalac SingletonObjectMainDemo.scala
scala SingletonObjectMainDemo
```

The `scalac` keyword invokes the compiler and generates the byte code for `SingletonObjectMainDemo`. The `scala` keyword is used to execute the byte code generated by compiler. The output is shown in Figure 1-19.

```
scala> object SingletonObjectMainDemo {
     |   def main(args: Array[String]) {
     |   println("This is printed in main method")
     |   }
     | }
defined object SingletonObjectMainDemo

scala> SingletonObjectMainDemo.main(Array(""))
This is printed in main method
```

Figure 1-19. *Calling main method of singleton object in REPL mode*

Companion Classes and Objects

An object with the same name as a class is called a companion object and the class is called a companion class.

The following is the code for companion objects. Save this code as `CompanionExample.scala` file and execute the program using this command.

```
scalac CompanionExample.scala
scala CompanionExample
```

```
//companion class
class Author(authorId:Int, authorName:String){
val id = authorId
val name = authorName
override def toString() =
this.id +" "+" , "+ this.name
}

//companion object
object Author{
def message(){
println("Welcome to Apress Publication")
}
```

```scala
def display(au:Author){
println("Author Details: " + au.id+","+au.name);
   }
}

object CompanionExample {
  def main(args: Array[String]) = {
      var author=new Author(1001,"Dharanidharan")
      Author.message()
      Author.display(author)
    }
}
```

The output of this program is shown in Figure 1-20.

```
c:\scala_programs>scalac CompanionExample.scala

c:\scala_programs>scala CompanionExample
Welcome to Apress Publication
Author Details: 1001,Dharanidharan
```

Figure 1-20. CompanionExample.scala output

Case Classes

Case classes are like the regular classes that are very useful for modeling immutable data. Case classes are useful in pattern matching, as we discuss later in this chapter. The keyword case class is used to create a case class. Here is the syntax for creating case classes:

```
case class <class_name> ( <variable 1>:<data_type>, <variable n>:
<data_type> )
```

The following code illustrates a case class.

```scala
scala> case class ApressBooks(
     | bookId:Int,
     | bookName:String,
```

```
| bookAuthor:String
| )
```

Figure 1-21 shows case class output.

```
scala> case class ApressBooks (
     | bookId : Int,
     | bookName : String,
     | bookAuthor : String
     | )
defined class ApressBooks
```

Figure 1-21. *Example for case class*

The case classes can be instantiated (object creation) without using the new keyword. All the case classes have an apply method by default that takes care of object creation. Refer to Figure 1-22.

```
scala> val book1 = ApressBooks(101,"Practical Spark","Subhashini Chellapan")
book1: ApressBooks = ApressBooks(101,Practical Spark,Subhashini Chellapan)

scala> val book2 = ApressBooks(102,"Practical Scala","Dharanitharan Ganesan")
book2: ApressBooks = ApressBooks(102,Practical Scala,Dharanitharan Ganesan)

scala> println(s"The book ${book1.bookName} is written by ${book1.bookAuthor}")
The book Practical Spark is written by Subhashini Chellapan

scala> println(s"The book ${book2.bookName} is written by ${book2.bookAuthor}")
The book Practical Scala is written by Dharanitharan Ganesan
```

Figure 1-22. *Case class object creation*

Case classes are compared by structure and not by reference (Figure 1-23).

```
scala> case class Authors(authorName:String,publisher:String)
defined class Authors

scala> val author1 = Authors("Dharanidharan", "Apress")
author1: Authors = Authors(Dharanidharan,Apress)

scala> val author2=Authors("Dharanidharan","Apress")
author2: Authors = Authors(Dharanidharan,Apress)

scala> author1 == author2
res0: Boolean = true
```

Figure 1-23. *Example for case class*

Even though author1 and author2 refer to different objects, the value of each object is equal.

Pattern Matching on Case Classes

Case classes are useful in pattern matching. In the following example, Books is an abstract superclass that has two concrete Book types implemented with case classes. Now we can do pattern matching on these case classes. The code is shown here and the results are displayed in Figure 1-24.

```
scala> abstract class Books
defined class Books

scala> case class ApressBooks(bookID:Int, bookName:String,
publisher:String) extends Books
defined class ApressBooks

scala> case class SpringerBooks(bookID:Int, bookName:String,
publisher:String) extends Books
defined class SpringerBooks

scala> def showBookDetails(book:Books) = {
     | book match {
     | case SpringerBooks(id,name,publisher) => s"The book ${name} is
       published by ${publisher}"
```

```
  | case ApressBooks(id,name,publisher) => s"The book ${name} is
    published by ${publisher}"
  | }
  | }
```

```
scala> abstract class Books
defined class Books

scala> case class ApressBooks(bookID:Int, bookName:String, publisher:String ) extends Books
defined class ApressBooks

scala> case class SpringerBooks(bookID:Int, bookName:String, publisher:String ) extends Books
defined class SpringerBooks

scala> def showBookDetails(book : Books) = {
    | book match {
    | case SpringerBooks(id,name,publisher) => s"The book ${name} is published by ${publisher}"
    | case ApressBooks(id,name,publisher) => s"The book ${name} is published by ${publisher}"
    | }
    | }
showBookDetails: (book: Books)String

scala> showBookDetails(ApressBooks(101,"Practical Spark","Apress Publications"))
res39: String = The book Practical Spark is published by Apress Publications

scala> showBookDetails(SpringerBooks(102,"Practical Scala","Springer Media"))
res40: String = The book Practical Scala is published by Springer Media
```

Figure 1-24. *Pattern matching on case class*

Note   The abstract class and `extends` keyword are like the same in Java. It is used here to represent the different book types (i.e., Apress Books & Springer Books as generic books), which makes the `showBookDetails` function able to accept any type of book as a parameter.

Scala Collections

The collections in Scala are the containers for some elements. They hold the arbitary number of elements of the same or different types based on the type of collection. There are two types of collections:

- Mutable collections.

- Immutable collections.

The contents or the reference of mutable collections can be changed, immutable collections cannot be changed. Table 1-1 explains the most commonly used collections with their descriptions.

Table 1-1. *Commonly Used Collections in Scala*

Collection	Description
List	Homogeneous collection of elements
Set	Collection of elements of same type with no duplicates
Map	Collection of key/value pairs
Tuple	Collection of elements of different type but fixed size
Option	Container for either zero or one element

The following code describes various collections.

```
val booksList = List("Spark","Scala","R Prog", "Spark")
val booksSet = Set("Spark","Scala","R Prog", "Spark")
val booksMap = Map(101 -> "Scala", 102 -> "Scala")
val booksTuple = new Tuple4(101,"Spark", "Subhashini","Apress")
```

Figure 1-25 depicts the creation of different collections.

```
scala> val booksList = List("Spark","Scala","R Prog", "Spark")
booksList: List[String] = List(Spark, Scala, R Prog, Spark)

scala> val booksSet = Set("Spark","Scala","R Prog","Spark")
booksSet: scala.collection.immutable.Set[String] = Set(Spark, Scala, R Prog)

scala> val booksMap = Map(101 -> "Scala", 102 -> "Scala")
booksMap: scala.collection.immutable.Map[Int,String] = Map(101 -> Scala, 102 -> Scala)

scala> val booksTuple = new Tuple4(101, "Spark", "Subhashini" , "Apress")
booksTuple: (Int, String, String, String) = (101,Spark,Subhashini,Apress)
```

Figure 1-25. *Commonly used collections in Scala*

In Scala, the Option[T] is a container for either zero or one element of a given type. The Option can either be Some[T] or None[T], where T can be any given type. For example, Some is referred for any available value and None is reffered for no value (i.e., like null).

Scala Map always returns the value as Some[<given_type>] if the key is present and None if the key is not present. Refer to the following code and Figure 1-26.

```
val booksMap = Map(101 -> "Scala", 102 -> "Scala")
```

```
scala> val booksMap = Map(101 -> "Scala", 102 -> "Scala")
booksMap: scala.collection.immutable.Map[Int,String] = Map(101 -> Scala, 102 -> Scala)

scala> booksMap.get(101)
res53: Option[String] = Some(Scala)

scala> booksMap.get(105)
res54: Option[String] = None
```

Figure 1-26. *Example of Option[T] collection*

The getOrElse() method is used to get the value from an Option or any default value if the value is not present. Refer to Figure 1-27.

```
scala> booksMap.get(101).getOrElse("Book Not Available")
res55: String = Scala

scala> booksMap.get(105).getOrElse("Book Not Available")
res56: String = Book Not Available
```

Figure 1-27. *Example of getOrElse() method of Option[T] collection*

Iterating Over the Collection

The collections can be iterated using the iterator method. The iterator.hasNext method is used to find whether the collection has further elements and the iterator. next method is used to access the elements in a collection. The following code describes the iterator method and Figure 1-28 shows its output.

```
scala> val booksList = List("Spark","Scala","R Prog","Spark")
booksList: List[String] = List(Spark, Scala, R Prog, Spark)

scala> def iteratingList(booksList:List[String]){
     | val iterator = booksList.iterator
     | while(iterator.hasNext){
     | println(iterator.next)
     | }
     | }
```

```
scala> val booksList = List("Spark","Scala","R Prog", "Spark")
booksList: List[String] = List(Spark, Scala, R Prog, Spark)

scala> def iteratingList( booksList : List[String]){
     | val iterator = booksList.iterator
     | while (iterator.hasNext) {
     |    println(iterator.next)
     | }
     | }
iteratingList: (booksList: List[String])Unit

scala> iteratingList(booksList)
Spark
Scala
R Prog
Spark
```

Figure 1-28. *Iterating elements in the list*

Here is another example, for which output is shown in Figure 1-29.

```
scala> val booksMap = Map(101 -> "Scala", 102 -> "Scala")
booksMap: scala.collection.immutable.Map[Int,String] = Map(101 -> Scala,
102 -> Scala)

scala> def iteratingMap(booksMap:Map[Int,String]){
     | val iterator = booksMap.keySet.iterator
     | while(iterator.hasNext){
     | var key =iterator.next
     | println(s"Book Id:$key,BookName:{booksMap.get(key)}")
     | }
     | }
iteratingMap: (booksMap: Map[Int,String])Unit
```

```
scala> val booksMap = Map(101 -> "Scala", 102 -> "Scala")
booksMap: scala.collection.immutable.Map[Int,String] = Map(101 -> Scala, 102 -> Scala)

scala> def iteratingMap ( booksMap : Map[Int,String] ){
     | val iterator = booksMap.keySet.iterator
     | while(iterator.hasNext){
     | var key = iterator.next
     | println(s"Book Id :$key, BookName:${booksMap.get(key)}")
     | }
     | }
iteratingMap: (booksMap: Map[Int,String])Unit

scala> iteratingMap(booksMap)
Book Id :101, BookName:Some(Scala)
Book Id :102, BookName:Some(Scala)
```

Figure 1-29. *Iterating elements in the Map*

Common Methods of Collection

The following are the common frequently used methods on various available collections.

- `filter`
- `map`
- `flatMap`
- `distinct`
- `foreach`

Figure 1-30 shows an illustration of commonly used methods on different collections.

```
scala> val booksList=List("Spark","Scala","R Prog","Spark")
booksList: List[String] = List(Spark, Scala, R Prog, Spark)

scala> val booksSet=List("Spark","Scala","R Prog","Spark")
booksSet: List[String] = List(Spark, Scala, R Prog, Spark)

scala> booksList.distinct
res5: List[String] = List(Spark, Scala, R Prog)

scala> booksList.foreach(println)
Spark
Scala
R Prog
Spark

scala> booksSet.map(name => s"The book name is $name").foreach(println)
The book name is Spark
The book name is Scala
The book name is R Prog
The book name is Spark

scala> booksSet.filter(name => name.equals("Spark")).foreach(println)
Spark
Spark

scala>
```

Figure 1-30. *Commonly used methods of collections*

The function { name => name.equals("Spark") } used inside the filter method is
called as an anonymous function, is discussed later in this chapter.
The flatMap unwraps all the elements of the collection inside a collection and forms a
single collection as shown in Figure 1-31.

```
scala> val numbersList = List(List(1,2,3),List(4,5,6),List(7,8,9))
numbersList: List[List[Int]] = List(List(1, 2, 3), List(4, 5, 6), List(7, 8, 9))

scala> numbersList.flatMap(list => list)
res82: List[Int] = List(1, 2, 3, 4, 5, 6, 7, 8, 9)
```

Figure 1-31. *Commonly used operations - flatMap*

Functional Programming Aspects of Scala

Let us understand the functional programming aspects of Scala. Scala supports anonymous functions, higher order functions, function composition, function currying, nested functions, and functions with variable length parameters (see Figure 1-32).

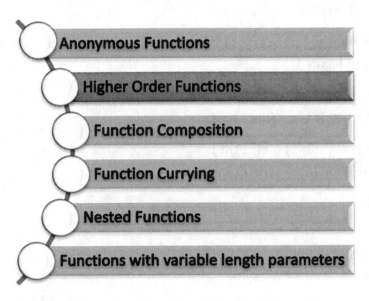

Figure 1-32. Functional programming aspects of Scala

Anonymous Functions

An anonymous function is a function that is not defined with a name and is created for single use. Like other functions, it also accepts input parameters and returns outputs. In simple words, these functions do not have a name but work like a function.

Anonymous functions can be created by using the => symbol and by _ (i.e., wildcard). They are also represented as lambda functions.

The function that follows can be used to calculate the sum of two numbers. It accepts two integers as input parameters and returns an integer as output.

```
def sumOfNumbers(a:Int,b:Int) : Int = {
return a + b
}
```

Calling this function using the defined name sumOfNumbers(2,3) returns the output 5. The anonymous function does not need the name to be defined explictly and because Scala has a strong built-in type inference mechanism, the data type need not be explicitly specified. Also, the return keyword can be ignored because the last statement in the function is a return statement by default. The same function can be written as

```
(a:Int, b:Int) => a+b
```

It can also be denoted as (_:Int)+(_:Int) using the _ wildcard character. Refer the following code and Figure 1-33.

```
scala> val sum = (a:Int, b:Int) => a+b
sum: (Int, Int) => Int = <function2>

scala> val diff = (_:Int) - (_:Int)
diff: (Int, Int) => Int = <function2>

scala> sum(3,2)
res12: Int = 5

scala> diff(3,2)
res13: Int = 1

scala> var sum = (a:Int,b:Int) => a+b
sum: (Int, Int) => Int = $$Lambda$1690/526318751@32ae8593

scala> var diff = (_:Int)-(_:Int)
diff: (Int, Int) => Int = $$Lambda$1692/90338335@7d82b654

scala> sum(3,2)
res85: Int = 5

scala> diff(3,2)
res86: Int = 1
```

Figure 1-33. *Anonymous functions*

Here, the anonymous function (a:Int,b:Int) => a+b is assigned to a variable as a value that proves that the function is also a value in functional programming. The left side is the input parameter to the function and the right side is the return value of the function.

Higher Order Functions

Functions that accept other functions as a paramenter or return a function are called higher order functions. The most common example of a higher order function in Scala is a map function applicable on collections.

If the function accepts another function as a parameter, the input parameter must be defined as shown in the following code and Figure 1-34.

```
scala> def normalFunc(inputString:String) = {
     | println(inputString)
     | }
normalFunc: (inputString: String)Unit

scala> def funcAsParameter(str:String,anyFunc:(String) => Unit) {
     | anyFunc(str)
     | }
funcAsParameter: (str: String, anyFunc: String => Unit)Unit

scala> funcAsParameter("This is a Higher order function",normalFunc)
This is a Higher order function
referenceName:(function_params) => returnType
```

```
scala> def normalFunc(inputString:String) = {
     |     println(inputString)
     | }
normalFunc: (inputString: String)Unit

scala> def funcAsParameter(str:String, anyFunc:(String)=>Unit) {
     |     anyFunc(str)
     | }
funcAsParameter: (str: String, anyFunc: String => Unit)Unit

scala> funcAsParameter("This is a Higher order fucntion", normalFunc)
This is a Higher order fucntion
```

Figure 1-34. *Higher order functions*

Here, the function funcAsParameter accepts another function as a parameter and returns the same function when it is called. Table 1-2 shows input parameter and return types of higher order functions.

Table 1-2. *Listing Higher Order Function Types*

Input Parameter	Return Type
Value	Function
Function	Function
Function	Value

Function Composition

In Scala, multiple functions can be composed together while calling. This is known as function composition. Refer to the following code and Figure 1-35.

```scala
scala> def concatValues(str1:String,str2:String):String = {
     | var concatedValue = str1.concat(str2);
     | concatedValue
     | }
concatValues: (str1: String, str2: String)String

scala> def display(dispValue:String) = {
     | print(dispValue)
     | }
display: (dispValue: String)Unit

scala> display(concatValues("Practical","Spark"))
PracticalSpark
```

```scala
scala> def concatValues(str1: String, str2: String): String = {
     |     var concatedValue = str1.concat(str2);
     |     concatedValue
     | }
concatValues: (str1: String, str2: String)String

scala> def display(dispValue: String) = {
     |     print(dispValue)
     | }
display: (dispValue: String)Unit

scala> display(concatValues("Practical", "Spark"))
PracticalSpark
```

Figure 1-35. *Function composition*

Here, the functions display and concatValues are composed together while calling.

Function Currying

The process of transforming a function that takes multiple arguments as parameters into a function with a single argument as a parameter is called *currying*. Function currying is primarily used to create a partially applied function. Partially applied functions are used to reuse a function invocation or to retain some of the parameters. In such cases, the number of parameters must be grouped as parameter lists. A single function can have multiple parameter lists, as shown here:

```
def multiParameterList(param1:Int)(param2:Int,param3:String){
    println("This function has two parameter lists")
    println("1st parameter list has single parameter")
    println("2nd parameter list has two parameters")
}
```

The following code and Figure 1-36 represent a function without currying.

```
scala> def bookDetails(id:Int)(name:String)(author:String){
     | println("The book id is " + id)
     | println("The book name is "+name)
     | println("The book author is "+author)
     | }
```

```
scala> def bookDetails(id:Int)(name:String)(author:String) {
     | println("The book id is "+id)
     | println("The book name is "+name)
     | println("The book author is "+author)
     | }
bookDetails: (id: Int)(name: String)(author: String)Unit

scala> bookDetails(101)("Practical Spark")("Dharanitharan G")
The book id is 101
The book name is Practical Spark
The book author is Dharanitharan G
```

Figure 1-36. *Without function currying*

When a function is called with fewer parameter lists, it yields a partially applied function, as illustrated in Figure 1-37.

```
scala> var newBookDetails=bookDetails(101)("Practical Spark")_
newBookDetails: String => Unit = $$Lambda$1695/150600414@310ba6ea
```

Figure 1-37. *Partially applied function: Function currying*

The bookDetails function is called by passing a lesser number of parameter lists than its total number of parameter lists. This can be done by simply using _ instead of a parameter list (see Figure 1-38).

```
scala> var newBookDetails=bookDetails(101)("Practical Spark")_
newBookDetails: String => Unit = $$Lambda$1695/150600414@310ba6ea

scala> newBookDetails("Dharanitharan G")
The book id is 101
The book name is Practical Spark
The book author is Dharanitharan G
```

Figure 1-38. *Function currying*

Nested Functions

Scala allows the user to define functions inside a function. This is known as nested functions, and the inner function is called a local function. The following code and Figure 1-39 represent the nested function.

```
scala> def bookAssignAndDisplay(bookId:Int,bookname:String) = {
    | def getBookDetails(bookId:Int,bookName:String):String = {
    | s"The bookId is $bookId and book name is $bookName"
    | }
    | def display{
    | println(getBookDetails(bookId,bookName))
    | }
    | display
    | }
```

```
scala>    def bookAssignAndDisplay(bookId:Int, bookname: String) = {
     |        def getBookDetails(bookId:Int, bookName: String): String = {
     |          s"The bookId is $bookId and book name is $bookName"
     |        }
     |        def display {
     |          println(getBookDetails(bookId,bookName))
     |        }
     |        display
     |     }
bookAssignAndDisplay: (bookId: Int, bookname: String)Unit

scala> bookAssignAndDisplay(101,"Practical Spark")
The bookId is 101 and book name is Practical Spark
```

Figure 1-39. *Nested functions*

Here, two inner functions are defined inside the function bookAssignAndDisplay.
The getBookDetails and display are the inner functions. The following code and
Figures 1-40 and 1-41 show the scope of the outer function.

```
scala> def outerFunction(){
     | var outerVariable ="Out"
     | def innerFunction(){
     | println(s"The value of outerVariable is : $outerVariable")
     | }
     | innerFunction()
     | }
outerFunction: ()Unit

scala> def outerFunction(){
     |     var outerVariable = "Out"
     |     def innerFunction(){
     |       var innerVariable = "In"
     | println(s"The value of outerVariable is : $outerVariable")
     |     }
     |     innerFunction()
     | }
outerFunction: ()Unit

scala> outerFunction()
The value of outerVariable is : Out
```

Figure 1-40. *Scope of outer function variable*

```scala
scala> def outerFunction(){
     | var outerVariable = "Out"
     | def innerFunction(){
     | var innerVariable ="In"
     | println(s"The value of outerVariable is :$outerVariable")
     | }
     | innerFunction()
     | println(s"The value of innerVariable is :$innerVariable")
     | }
```

```scala
scala> def outerFunction(){
     |     var outerVariable = "Out"
     |     def innerFunction(){
     |       var innerVariable = "In"
     |       println(s"The value of outerVariable is : $outerVariable")
     |     }
     | innerFunction()
     | println(s"The value of innerVariable is : $innerVariable")
     | }
<console>:20: error: not found: value innerVariable
       println(s"The value of innerVariable is : $innerVariable")
                                                  ^
```

Figure 1-41.   *Scope of inner function variable*

The variables declared in the outer function can be accessed in the inner function, but the variables declared in the inner function do not have the scope in the outer function.

Functions with Variable Length Parameters

The variable length parameters allow passing any number of arguments of the same type to the function when it is called. The following code represents the functions with variable length parameters. Figure 1-42 displays the output.

```scala
scala> def add(values:Int*)={
     | var sum =0;
     | for (value <- values){
     | sum = sum+value
     | }
```

```
        | sum
        | }
add: (values: Int*)Int

scala> def add(values: Int*) = {
        |       var sum = 0;
        |       for (value <- values){
        |          sum =sum+value
        |       }
        |       sum
        | }
add: (values: Int*)Int

scala> var sum = add(1, 2, 3, 4, 5, 6, 7, 8, 9);
sum: Int = 45

scala> println(s"The sum of all arguments is $sum");
The sum of all arguments is 45
```

Figure 1-42. *Variable length parameters*

The variable length parameters can be defined using the * operator. When it is defined as Int*, it is mandatory to pass all parameters as Int. It is possible to pass other parameters along with variable length parameters but the variable length parameters should be the last in the parameter list. The following code and Figure 1-43 show variable length parameters with other parameters.

```
scala> def add(ops:String,values:Int*) = {
    | println(s"Performing $ops of all elements in variable length
      parameter")
    | var sum = 0;
    | for(value <- values){
    | sum =sum+value
    | }
    | sum
    | }
add: (ops: String, values: Int*)Int
```

```
scala> def add(ops:String,values: Int*) = {
     | println(s"Performing $ops of all elements in variable lenth parameter")
     | var sum = 0;
     |     for (value <- values){
     |        sum =sum+value
     |     }
     |     sum
     | }
add: (ops: String, values: Int*)Int

scala> var sum = add("addition",1, 2, 3, 4, 5, 6, 7, 8, 9);
Performing addition of all elements in variable lenth parameter
sum: Int = 45

scala> println(s"The sum of all arguments is $sum");
The sum of all arguments is 45
```

Figure 1-43. *Variable length parameters with other parameters*

A function cannot accept two variable length parameters, as reflected in Figure 1-44.

```
scala> def add(ops:String,values: Int*,values2: Int*) = {
     | println(s"Performing $ops of all elements in variable lenth parameter")
     |     var sum = 0;
     |     for (value <- values){
     |        sum =sum+value
     |     }
     |     sum
     | }
<console>:12: error: *-parameter must come last
       def add(ops:String,values: Int*,values2: Int*) = {
```

Figure 1-44. *Error: Multiple variable length parameters*

Note   In Scala, there are no primitive data types. Everthing is an object.

Scala doesn't have operators. The operators are known as methods. Hence, we can not use * while importing a package.

The packages can be imported as shown here:

```
import java.io._
```

Reference Links

- https://docs.scala-lang.org/tour/tour-of-scala.html

Points to Remember

- A function that has no side effects is called a pure function.

- In Scala, the compiler can identify the type of variable based on the initialization of the variable by the built-in type inference mechanism.

- The lazy evaluation feature allows the user to defer the execution of any expression until it is needed using the lazy keyword.

- A Scala class cannot have static variables and methods. Instead, Scala classes can have singleton objects or companion objects.

- Case classes are like the regular classes that are very useful for modeling immutable data.

- An anonymous function is a function that is not defined with a name and is created for a single use.

- Functions that accept other functions as a parameter or return a function are called higher order functions.

In next chapter, we discuss the installation and cluster setup for Apache Spark.

CHAPTER 2

Single and Multinode Cluster Setup

This chapter explains how to install Apache Spark on a single and multinode cluster. The recommended background for this chapter is to have some prior experience with basic Unix commands.

By end of this chapter, you will be able to do the following:

- Set up a single and multinode spark cluster.

- Understand the various configurations in the Spark cluster setup.

- Perform basic administration activities on the Spark cluster.

Note We recommend following the step-by-step instructions and the complete procedure to create single-node and multinode Spark clusters based on the requirements.

Spark Multinode Cluster Setup

Follow this guide to create a three-node spark cluster.

Recommended Platform

We recommend following this procedure to complete the cluster setup with minimal operating system requirements for learning purposes.

© Subhashini Chellappan, Dharanitharan Ganesan 2018
S. Chellappan and D. Ganesan, *Practical Apache Spark*, https://doi.org/10.1007/978-1-4842-3652-9_2

Operating System

Windows is supported as a development platform, but Linux is recommended for the development and deployment cluster. We recommend using Ubuntu - 14.0/16.0 or later. You can download the Ubuntu iso file from `http://releases.ubuntu.com/trusty/`. (Please note that this link might be changed in future as it depends on the Ubuntu release team.)

Because we follow the steps to create a three-node cluster, we need three Ubuntu machines that can be created in any cloud service provider or on-premises nodes.

To create the cluster on your personal PC, we recommend using any virtual machine provider like Oracle VirtualBox or VM Workstation to create multiple machines.

We have used Oracle VirtualBox to create three Ubuntu virtual machines and the details are given next (see Figure 2-1).

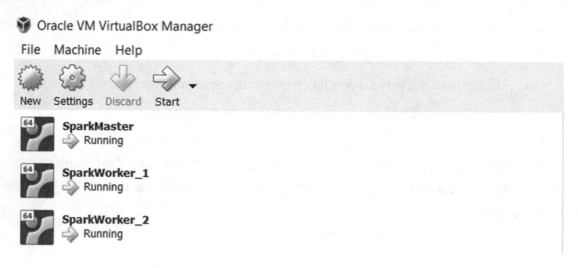

Figure 2-1. *Oracle VM VirtualBox Manager*

Follow the steps given here to install VirtualBox and create the virtual machines. Download the latest version of Oracle VirtualBox from `https://download.virtualbox.org/virtualbox/5.2.18/VirtualBox-5.2.18-124319-Win.exe`. (Please note that this link might be changed in the future as it depends on the Oracle release team.) Once the application is downloaded, right-click the application and run it as administrator (see Figure 2-2).

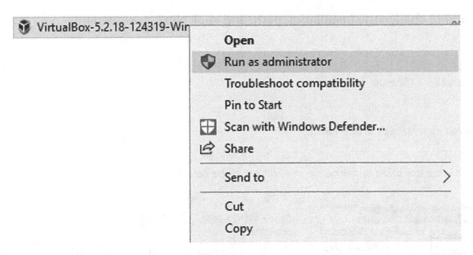

Figure 2-2. *Starting the VirtualBox installation*

Click Next on the screen shown in Figure 2-3 to proceed with the installation.

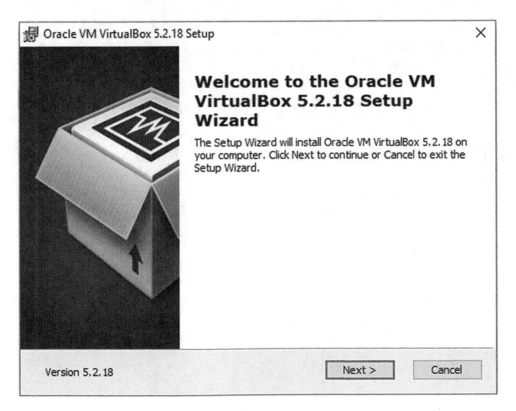

Figure 2-3. *VirtualBox installation step*

Browse the location to change the installation path, and click Next, as shown in Figure 2-4.

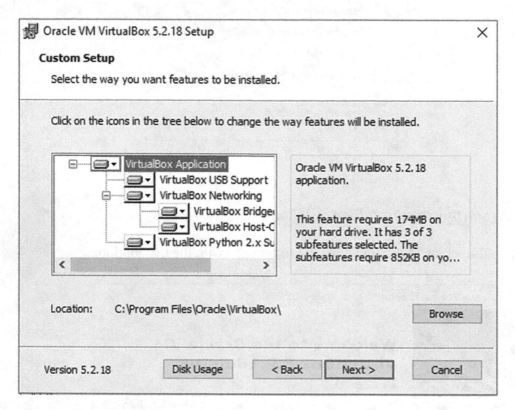

Figure 2-4. *VirtualBox installation continued*

Next, select the required features for custom installation, as shown in Figure 2-5.

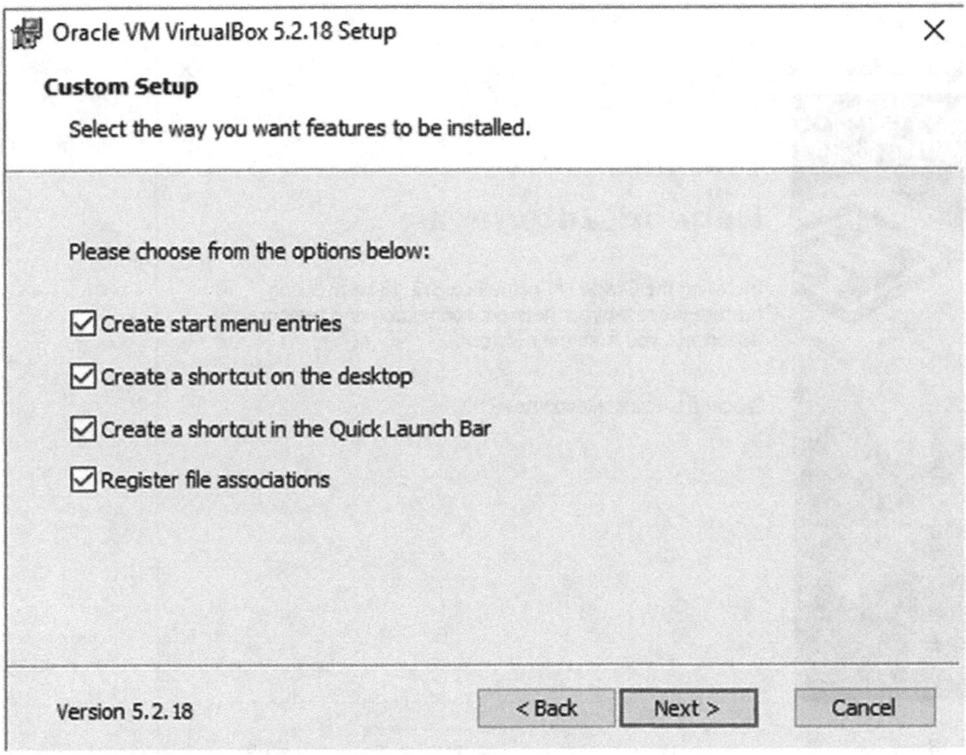

Figure 2-5. *VirtualBox installation steps continued*

Select Yes in the next step, depicted in Figure 2-6, to install the VirtualBox network interfaces.

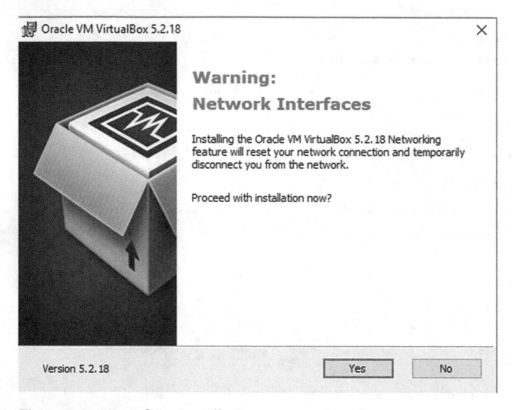

Figure 2-6. *VirtualBox installation steps continued*

Click Install on the next wizard page, shown in Figure 2-7, to proceed with the installation.

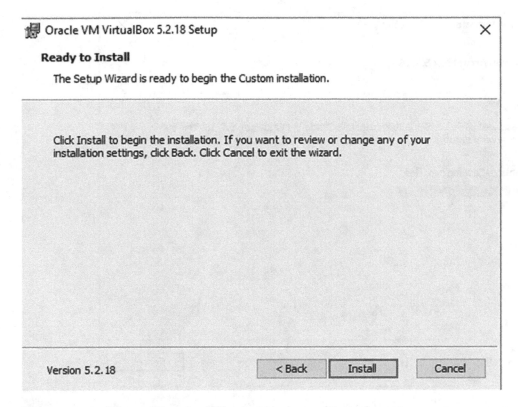

Figure 2-7. *VirtualBox installation steps continued*

Wait for a few minutes for the installation to complete. You will see the wizard page shown in Figure 2-8.

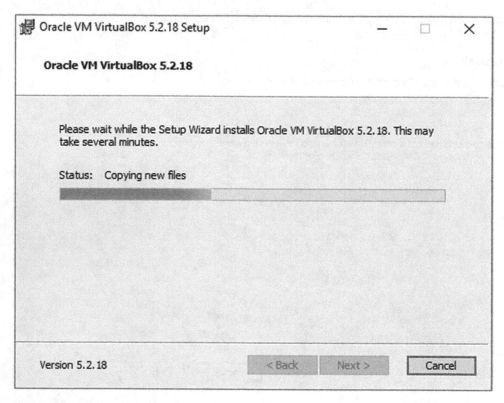

Figure 2-8. *VirtualBox installation steps continued*

Click Finish, as indicated in Figure 2-9, to complete the installation process.

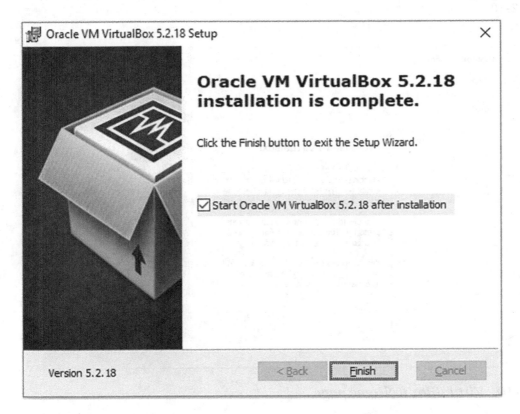

Figure 2-9. *VirtualBox installation steps continued*

Once the installation is complete, start VirtualBox. You will see the welcome page shown in Figure 2-10.

Figure 2-10. *VirtualBox welcome page*

To create the virtual machines, first click New to create a new virtual machine. Specify the name of the virtual machine and select Linux as the type of operating system. Click next. These steps are shown in Figure 2-11.

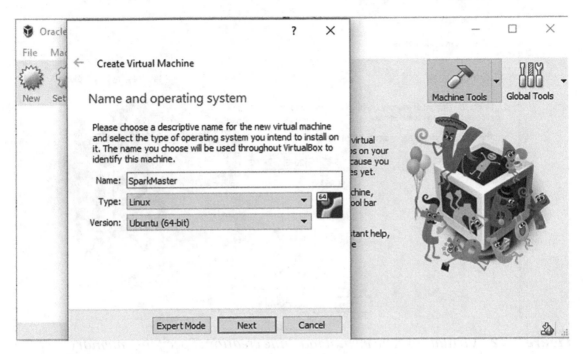

Figure 2-11. *VirtualBox new virtual machine creation*

Specify the memory to be allotted to the virtual machine in the dialog box shown in Figure 2-12. The recommended memory is 1024 MB or more. When complete, click Next.

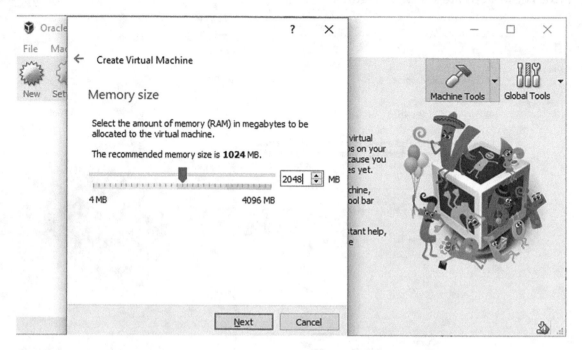

Figure 2-12. *VirtualBox new virtual machine creation: Specifying memory*

Select Create A Virtual Hard Disk Now from the available options in the Hard Disk dialog box (see Figure 2-13) to create the new virtual hard disk for the machine. Click Next.

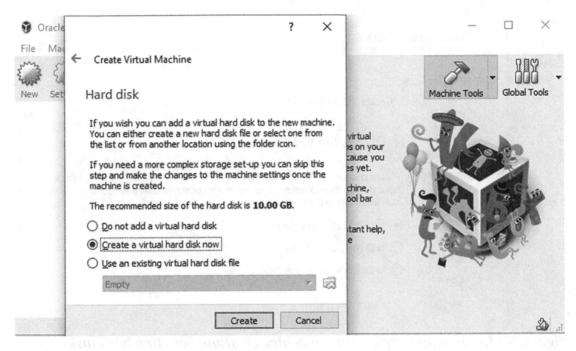

Figure 2-13. *VirtualBox new virtual machine creation: Creating a hard disk*

Select VDI (VirtualBox Disk Image) as the hard disk file type, as shown in Figure 2-14. Click next.

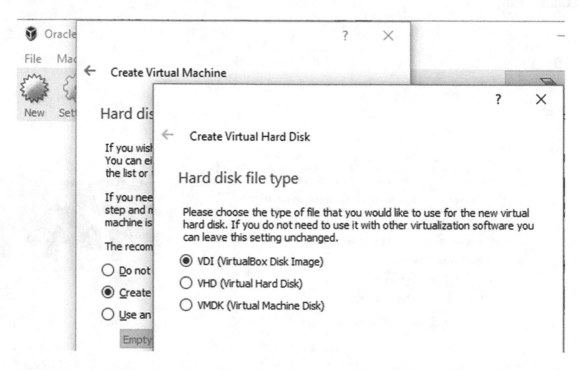

Figure 2-14. *VirtualBox new virtual machine creation: Selecting hard disk file type*

Select Dynamically Allocated as the virtual hard disk storage type (see Figure 2-15) to ensure the size of the hard disk grows dynamically. Click Next.

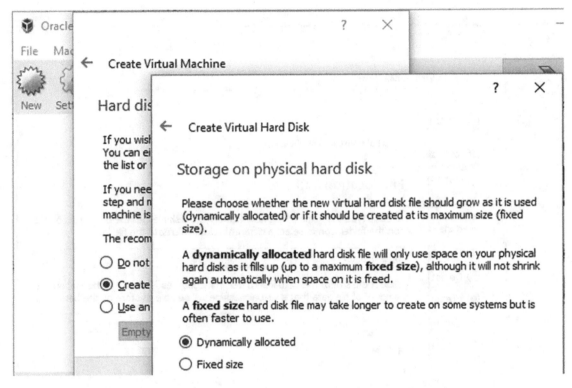

Figure 2-15. *VirtualBox new virtual machine creation: Selecting hard disk storage type*

Next, select the file location and the size of the virtual hard disk and click Create, as shown in Figure 2-16. This size is the limit on the amount of file data stored on the hard disk.

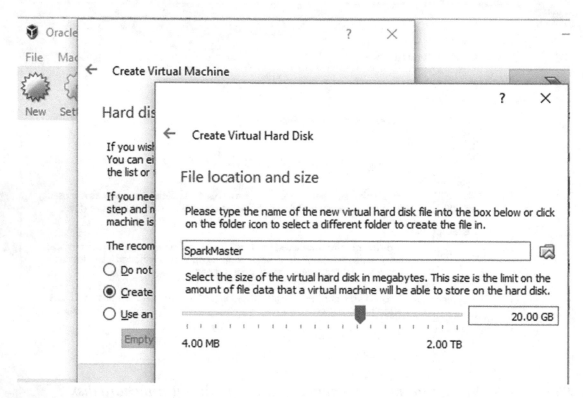

Figure 2-16. *VirtualBox new virtual machine creation: Specifying hard disk file location and size*

Once the machine is created as shown in Figure 2-17, click Settings to specify the iso file to install Ubuntu in the created virtual machine and also to change the network settings.

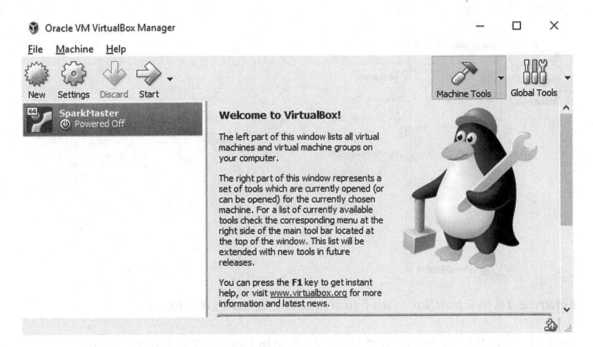

Figure 2-17. *VirtualBox new virtual machine*

In the Settings dialog box, select Storage and click the Adds Optical Drive icon as shown in Figure 2-18 to specify the Ubuntu iso file.

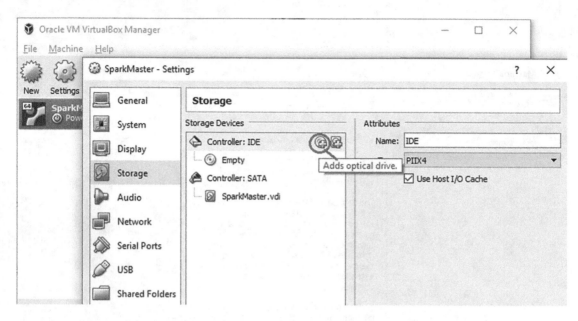

Figure 2-18. *VirtualBox new virtual machine Settings dialog box*

Click Choose Disk, as displayed in Figure 2-19, to browse for the Ubuntu iso file.

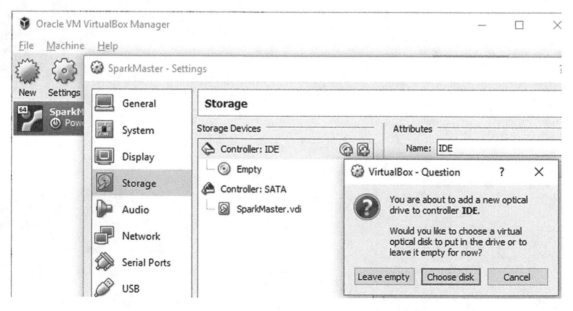

Figure 2-19. *VirtualBox new virtual machine: Choosing the iso disk file*

Select the downloaded iso file and click OK. Now select Network in the Settings dialog box to change the network settings. Refer to Figure 2-20.

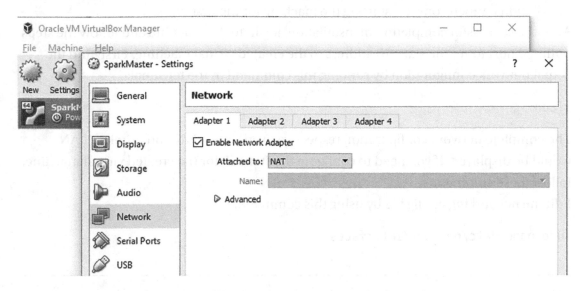

Figure 2-20. *VirtualBox new virtual machine: Network settings*

Select the bridged adapter as shown in Figure 2-21 and click OK to complete the network settings.

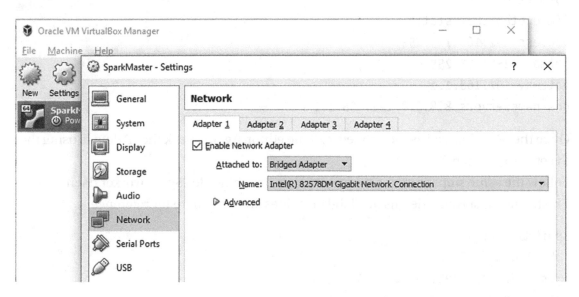

Figure 2-21. *VirtualBox new virtual machine: Network adapter selection*

Once these settings are completed, start the virtual machine to complete the installation. Specify the required username and password during the installation procedure. The installation happens only on starting the machine for the first time.

After the successful completion of installation, log in to the machine and follow the steps outlined next to set the static IP address to the created virtual machine. First, find the actual network configuration by running this command in the terminal:

```
ifconfig
```

The complete network configuration respective to ethernet LAN and Wireless LAN would be displayed. If you need to configure the static IP for the created virtual machine, follow these commands.

Edit the networking config file by using this command:

```
sudo nano /etc/network/interfaces
```

Note   You can use either the nano editor or the vi editor.

Change the details as shown here.

```
auto eth0
iface eth0 inet static
address 192.168.163.153
netmask 255.255.255.0
gateway 192.168.1.1
dns-nameservers 8.8.8.8 192.168.1.1
```

Once the IP is changed as mentioned, exit the editor. Again check the IP details using the `ifconfig` command.

Follow the same steps to create two more virtual machines for Spark worker daemons. The IP and Host Name details of all the machines are given here for reference.

```
SparkMaster:
  Host Name - SparkMaster
  Ip Address - 192.168.163.153
SparkWorker_1:
  Host Name - SparkWorker1
```

```
  Ip Address - 192.168.163.152
SparkWorker_2:
  Host Name - SparkWorker2
  Ip Address - 192.168.163.151
```

Log in to all the machines with the username and password. We have used Putty to log in to the machines through SSH (i.e., Secured Shell). The reference image for the SparkMaster machine given in Figure 2-22. Note that we have used vagrant as the username.

```
vagrant@SparkMaster: ~
login as: vagrant
vagrant@192.168.163.153's password:
Welcome to Ubuntu 12.04 LTS (GNU/Linux 3.2.0-23-generic x86_64)

 * Documentation:  https://help.ubuntu.com/
New release '14.04.3 LTS' available.
Run 'do-release-upgrade' to upgrade to it.

Welcome to your Vagrant-built virtual machine.
Last login: Wed May 16 09:49:59 2018 from 192.168.163.1
vagrant@SparkMaster:~$
```

Figure 2-22.   *Login reference for SparkMaster machine*

Figure 2-23 shows the login reference for the SparkWorker1 machine.

```
vagrant@SparkWorker1: ~
login as: vagrant
vagrant@192.168.163.152's password:
Welcome to Ubuntu 12.04 LTS (GNU/Linux 3.2.0-23-generic x86_64)

 * Documentation:  https://help.ubuntu.com/
New release '14.04.3 LTS' available.
Run 'do-release-upgrade' to upgrade to it.

Welcome to your Vagrant-built virtual machine.
Last login: Wed May 16 09:55:27 2018
vagrant@SparkWorker1:~$
```

Figure 2-23.   *Login reference for SparkWorker1 machine*

Figure 2-24 depicts the login reference for the SparkWorker2 machine.

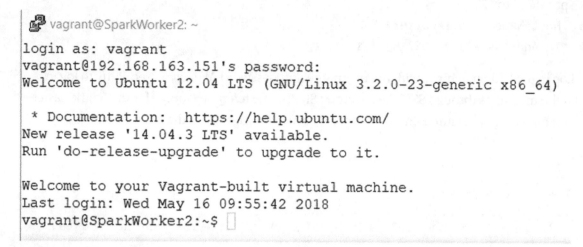

Figure 2-24. *Login reference for SparkWorker2 machine*

Add the hostnames of all the machines in the hosts file of each machine as shown here.

192.168.163.153 SparkMaster
192.168.163.152 SparkWorker1
192.168.163.151 SparkWorker2

The hosts file in each machine can be edited using the vi editor, as shown in Figure 2-25.

```
vagrant@SparkMaster:~$ sudo vi /etc/hosts
```

Figure 2-25. *Editing the host files to add the* IP *address*

We have used the latest version of Spark, 2.3.0. It can be downloaded from http://redrockdigimark.com/apachemirror/spark/spark-2.3.0/spark-2.3.0-bin-hadoop2.7.tgz. For the most up-to-date download, follow the Apache Spark documentation to download the required version.

Prerequisites

Ensure that Java 1.8 is installed on all machines. Execute this command to install Java on all the nodes:

```
sudo apt-get install  openjdk-8-jdk
```

If the openjdk-8-jdk package is not available for the Ubuntu version you are using, download the jdk package from this link and extract the tar in all the machines: http://download.oracle.com/otn-pub/java/jdk/8u172-b11/a58eab1ec242421181065cdc37240b08/jdk-8u172-linux-x64.tar.gz.

Note   The given link is subject to change. Visit http://download.oracle.com/ and check for available downloads and updates.

Ensure the proper installation of Java using the java -version command in all the three nodes, as shown in Figure 2-26.

```
vagrant@SparkMaster:~$ java -version
java version "1.8.0_77"
Java(TM) SE Runtime Environment (build 1.8.0_77-b03)
Java HotSpot(TM) 64-Bit Server VM (build 25.77-b03, mixed mode)
```

Figure 2-26. *Check the Java version*

Now, Set the JAVA_HOME in the .bashrc profile file of all the machines as shown in Figure 2-27. For example:

```
export JAVA_HOME=<your_java_installation_path>
export PATH=$PATH:$JAVA_HOME/bin
```

```
vagrant@SparkMaster:~$ sudo vi ~/.bashrc
```

Figure 2-27. *Editing the .bashrc file*

Add these lines to the end of the .bashrc file:

```
export JAVA_HOME=/home/vagrant/java8
export PATH=$PATH:$JAVA_HOME/bin
```

Note   The installation path could be different from /home/vagrant/java8.

Once the PATH is added, use

```
source ~/.bashrc
```

to update the .bashrc file to the same session without restarting the machine, as displayed in Figure 2-28.

```
vagrant@SparkMaster:~$ source ~/.bashrc
```

Figure 2-28. *Updating the .bashrc file*

Verify the updated path details, using the code shown in Figure 2-29.

```
vagrant@SparkMaster:~$ echo $JAVA_HOME
/home/vagrant/java8
```

Figure 2-29. *Verify the Java home path*

Spark Installation Steps

Download the Spark binaries in master node by using this Unix wget command:

```
sudo wget http://redrockdigimark.com/apachemirror/spark/spark-2.3.0/spark-2.3.0-bin-hadoop2.7.tgz
```

Copy the downloaded binaries to all the other nodes using the scp command in this code as shown in Figure 2-30.

```
scp spark-2.3.0-bin-hadoop2.7.tgz vagrant@SparkWorker1:/home/vagrant
```

```
vagrant@SparkMaster:~$ scp spark-2.3.0-bin-hadoop2.7.tgz  vagrant@SparkWorker1:/home/vagrant
spark-2.3.0-bin-hadoop2.7.tgz                                                          100%
vagrant@SparkMaster:~$ scp spark-2.3.0-bin-hadoop2.7.tgz  vagrant@SparkWorker2:/home/vagrant
spark-2.3.0-bin-hadoop2.7.tgz                                                          100%
vagrant@SparkMaster:~$
```

Figure 2-30. *Copy Spark binaries to other virtual machines*

Extract the .tgz zip file and rename the directory spark-2.3.0 in all three nodes, as shown here and in Figures 2-31 and 2-32.

tar - xvf spark-2.3.0-bin-hadoop2.7.tgz

```
vagrant@SparkMaster:~$ tar -xvf spark-2.3.0-bin-hadoop2.7.tgz
```

Figure 2-31. *Extracting the .tgz zip file*

mv spark-2.3.0-bin-hadoop2.7 spark-2.3.0

```
vagrant@SparkMaster:~$ mv spark-2.3.0-bin-hadoop2.7 spark-2.3.0
vagrant@SparkMaster:~$ ls
postinstall.sh    spark-2.3.0    spark-2.3.0-bin-hadoop2.7.tgz
```

Figure 2-32. *Unzipping the Spark binaries*

Now, set the SPARK_HOME in the .bashrc profile file of all the machines as shown here and in Figure 2-33.

export SPARK_HOME=<your_spark_installation_path>
export PATH=$PATH:$SPARK_HOME/bin:$SPARK_HOME/sbin

```
vagrant@SparkMaster:~$ sudo vi ~/.bashrc
```

Figure 2-33. *Verify .bashrc in other virtual machines*

Add the following lines to the end of the `.bashrc` file:

```
export SPARK_HOME=/home/vagrant/spark-2.3.0
export PATH=$PATH:$JAVA_HOME/bin
```

Note   The installation path could be different from `/home/vagrant/spark-2.3.0`.

Once the PATH is added, use `source ~/.bashrc` to update the `.bashrc` file to the same session without restarting the machine, as shown in Figure 2-34.

```
vagrant@SparkMaster:~$ source ~/.bashrc
```

Figure 2-34. *Updating the* `.bashrc` *file*

Verify the updated path details using the code shown in Figure 2-35.

```
vagrant@SparkMaster:~$ echo $SPARK_HOME
/home/vagrant/spark-2.3.0
```

Figure 2-35. *Verifying the Spark installation home path*

After installing Spark in all the nodes and the PATH variable is updated, specify the slave details in all the nodes (i.e., the worker node details for the Spark cluster) by following these steps in all three nodes.

1. Navigate to the `conf` directory in the Spark installation folder:

    ```
    cd /home/vagrant/spark-2.3.0/conf
    ```

2. Rename the `slaves.template` file `slaves`:

    ```
    mv slaves.template slaves
    ```

3. Edit the `slaves` file and add `SparkWorker1` and `SparkWorker2` at the end of the file (see Figure 2-36):

    ```
    vi slaves
    ```

```
vagrant@SparkMaster:~$ cd ~/spark-2.3.0/
vagrant@SparkMaster:~/spark-2.3.0$ cd conf/
vagrant@SparkMaster:~/spark-2.3.0/conf$ ls
docker.properties.template  metrics.properties.template  spark-env.sh.template
fairscheduler.xml.template  slaves.template
log4j.properties.template   spark-defaults.conf.template
vagrant@SparkMaster:~/spark-2.3.0/conf$ mv slaves.template slaves
vagrant@SparkMaster:~/spark-2.3.0/conf$ vi slaves █
```

Figure 2-36. *Add Spark master and worker details*

Add these lines to the slaves file:

```
SparkWorker1
SparkWorker2
```

4. Rename the spark-env.sh.template file spark-env.sh:

   ```
   mv spark-env.sh.template spark-env.sh
   ```

5. Edit the spark-env.sh file and add the JAVA_HOME path to the file:

   ```
   vi spark-env.sh
   ```

6. Add this line in the spark-env.sh file.

   ```
   export JAVA_HOME=/home/vagrant/java8
   ```

Check for the running services in all the three nodes (see Figure 2-37) to confirm that the Spark cluster is not running.

```
vagrant@SparkMaster:~$ jps
1762 Jps

vagrant@SparkWorker1:~$ jps
1742 Jps

vagrant@SparkWorker2:~$ jps
1872 Jps
```

Figure 2-37. *Checking running Java processes in all virtual machines*

Now, on the SparkMaster machine, start the processes by calling the script `start-all.sh` (see Figure 2-38), which starts the master process in SparkMaster and worker processes in both SparkWorker1 and SparkWorker2.

Also, the master process can be started using `start-master.sh` and worker processes can be started using `start-slaves.sh` separately.

```
vagrant@SparkMaster:~$ start-all.sh
starting org.apache.spark.deploy.master.Master
SparkWorker2: starting org.apache.spark.deploy.worker.Worker
SparkWorker1: starting org.apache.spark.deploy.worker.Worker
```

Figure 2-38. *Start the Spark master*

Now check for the running services in all three nodes as shown in Figure 2-39 to verify that the Spark cluster is running.

```
vagrant@SparkMaster:~$ jps
2209 Jps
2145 Master

vagrant@SparkWorker1:~$ jps
2258 Worker
2303 Jps

vagrant@SparkWorker2:~$ jps
2546 Jps
2500 Worker
```

Figure 2-39. *Checking running Spark process in all virtual machines*

Now, the three-node Spark cluster is running with master on the SparkMaster machine and worker on the SparkWorker1 and SparkWorker2 machines.

Spark Web UI

Let's go through the Spark Master user interface (UI) and Spark application UI in the following sections.

Spark Master UI

When the Spark cluster is running, browse the Spark UI using the following URL to learn about worker nodes attached to the master, running applications, and the cluster resources.

```
http://mastermachine-ip_address:8080/
```

```
Example: http://192.168.163.153:8080/
```

```
The following information can be found in the Spark Master Web UI
(see Figure 2-40):
```

```
URL: spark://SparkMaster:7077
REST URL: spark://SparkMaster:6066 (cluster mode)
Alive Workers: 2
Cores in use: 4 Total, O Used
Memory in use: 2.0 GB Total, 0.0 B Used
Applications: O Running, O Completed
Drivers: O Running, O Completed
Status: ALIVE
```

Also, the Workers, Running Applications, and Completed Application details would be updated in the same UI.

← → C   ⓘ 192.168.163.153:8080

 2.3.0   **Spark Master at spark://SparkMaster:7077**

URL: spark://SparkMaster:7077
REST URL: spark://SparkMaster:6066 *(cluster mode)*
Alive Workers: 2
Cores in use: 4 Total, 0 Used
Memory in use: 2.0 GB Total, 0.0 B Used
Applications: 0 Running, 0 Completed
Drivers: 0 Running, 0 Completed
Status: ALIVE

Figure 2-40. *Spark Master*

Figure 2-41 shows the Spark Web UI.

Workers (2)

Worker Id	Address
worker-20180516195830-192.168.163.151-56237	192.168.163.151:56237
worker-20180516195830-192.168.163.152-49344	192.168.163.152:49344

Running Applications (0)

Application ID	Name	Cores	Memory per Executor

Completed Applications (0)

Application ID	Name	Cores	Memory per Executor ▲

Figure 2-41. *Spark Web UI*

The worker node status, available cores, and the available memory are also updated as shown in Figure 2-42.

Workers (2)

Worker Id	Address	State	Cores	Memory
worker-20180516195830-192.168.163.151-56237	192.168.163.151:56237	ALIVE	2 (0 Used)	1024.0 MB (0.0 B Used)
worker-20180516195830-192.168.163.152-49344	192.168.163.152:49344	ALIVE	2 (0 Used)	1024.0 MB (0.0 B Used)

Figure 2-42. *Spark Web UI worker details*

Spark Application UI

When an application is submitted to the cluster, the browser and the Spark application UI need to know about the execution details and a list of tasks for the submitted job.

```
http://mastermachine-ip_address:4040/
```

For example, when the spark-shell is started it creates an application and then submits it to the cluster. The application would be in the running state until the spark-shell is closed.

Start the interactive shell by calling `spark-shell` in the $SPARK_HOME/bin directory (see Figure 2-43).

```
vagrant@SparkMaster:~$ spark-shell
2018-05-16 21:21:14 WARN  NativeCodeLoader:62 - Unable to load native-hadoop library
Setting default log level to "WARN".
To adjust logging level use sc.setLogLevel(newLevel). For SparkR, use setLogLevel(newLevel).
Spark context Web UI available at http://SparkMaster:4040
Spark context available as 'sc' (master = local[*], app id = local-1526505685727).
Spark session available as 'spark'.
Welcome to
      ____              __
     / __/__  ___ _____/ /__
    _\ \/ _ \/ _ `/ __/  '_/
   /___/ .__/\_,_/_/ /_/\_\   version 2.3.0
      /_/

Using Scala version 2.11.8 (Java HotSpot(TM) 64-Bit Server VM, Java 1.8.0_77)
Type in expressions to have them evaluated.
Type :help for more information.

scala>
```

Figure 2-43. *Starting spark-shell*

Note   spark-shell is an interactive shell that can be used for testing and debugging purposes.

Now, the spark application UI is available at http://192.168.163.153:4040/ as shown in Figure 2-44.

Spark Jobs (?)

User: vagrant
Total Uptime: 3.6 min
Scheduling Mode: FIFO

▸ Event Timeline

Figure 2-44. *Spark Web UI submitted jobs*

Stopping the Spark Cluster

In SparkMaster machine stops all the processes by calling the script `stop-all.sh`, which stops the master process in SparkMaster and worker processes in both SparkWorker1 and SparkWorker2 (see Figure 2-45).

```
vagrant@SparkMaster:~$ jps
2336 Jps
2145 Master
vagrant@SparkMaster:~$ stop-all.sh
SparkWorker2: stopping org.apache.spark.deploy.worker.Worker
SparkWorker1: stopping org.apache.spark.deploy.worker.Worker
stopping org.apache.spark.deploy.master.Master
vagrant@SparkMaster:~$ jps
2367 Jps
```

Figure 2-45. *Stopping the Spark processes*

Also, the master process can be stopped using `stop-master.sh` and worker processes can be stopped using `stop-slaves.sh` separately.

Spark Single-Node Cluster Setup

Follow this procedure to set up a Spark single-node cluster.

First, create the Ubuntu machine to install Spark, and run master and worker processes in the same machine that forms the single-node cluster (see Figure 2-46).

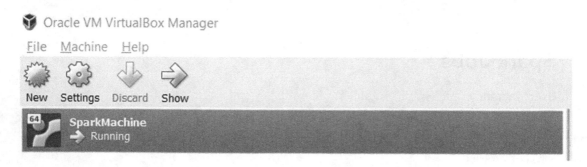

Figure 2-46. *VirtualBox virtual machine (master machine)*

The IP and hostname details of the machine are given here for reference.

```
SparkMachine:
  Host Name - SparkMachine
  Ip Address - 192.168.163.153
```

Log in to the machine with the username and password. We use Putty to log in to the machine through SSH. The reference image is given in Figure 2-47.

```
vagrant@SparkMachine: ~

login as: vagrant
vagrant@192.168.163.153's password:
Welcome to Ubuntu 12.04 LTS (GNU/Linux 3.2.0-23-generic x86_64)

 * Documentation:  https://help.ubuntu.com/
New release '14.04.3 LTS' available.
Run 'do-release-upgrade' to upgrade to it.

Welcome to your Vagrant-built virtual machine.
Last login: Mon May 21 05:37:54 2018
vagrant@SparkMachine:~$
```

Figure 2-47. *SparkMaster virtual machine login reference*

We are using the latest version of Spark, 2.3.0, which can be downloaded from http://redrockdigimark.com/apachemirror/spark/spark-2.3.0/spark-2.3.0-bin-hadoop2.7.tgz.
This link is subject to change. You should follow the Apache Spark documentation to download the required and most up-to-date version.

Prerequisites

Ensure that Java 1.8 is installed on all the machines.
Execute this command to install Java on all the nodes:

```
sudo apt-get install  openjdk-8-jdk
```

If the openjdk-8-jdk package is not available for the Ubuntu version you are using, download the jdk package from the following link and extract the tar in all the machines.

http://download.oracle.com/otn-pub/java/jdk/8u172-b11/
a58eab1ec242421181065cdc37240b08/jdk-8u172-linux-x64.tar.gz

Note This link might be modified over the time. Visit http://download.oracle.com/ and check for available downloads.

Ensure the proper installation of Java using the java -version command (see Figure 2-48).

```
vagrant@SparkMachine:~$ java -version
java version "1.8.0_77"
Java(TM) SE Runtime Environment (build 1.8.0_77-b03)
Java HotSpot(TM) 64-Bit Server VM (build 25.77-b03, mixed mode)
vagrant@SparkMachine:~$
```

Figure 2-48. *Verify Java version*

Next, set the JAVA_HOME in the .bashrc profile file. For example (see Figure 2-49):

export JAVA_HOME=<your_java_installation_path>
export PATH=$PATH:$JAVA_HOME/bin

```
vagrant@SparkMachine:~$ sudo vi ~/.bashrc
```

Figure 2-49. *Editing the .bashrc file*

Add these lines to the end of the .bashrc file:

export JAVA_HOME=/home/vagrant/java8
export PATH=$PATH:$JAVA_HOME/bin

Note The installation path could be different from /home/vagrant/java8.

Once the PATH is added, use `source ~/.bashrc` (see Figure 2-50) to update the `.bashrc` file to the same session without restarting the machine.

```
vagrant@SparkMachine:~$ source ~/.bashrc
```

Figure 2-50. *Updating the* `.bashrc` *file*

Verify the updated path details, as shown in Figure 2-51.

```
vagrant@SparkMachine:~$ echo $JAVA_HOME
/home/vagrant/java8
```

Figure 2-51. *Verify the Java home path*

Spark Installation Steps

Download the Spark binaries in the master node by using this Unix `wget` command:

```
sudo wget http://redrockdigimark.com/apachemirror/spark/spark-2.3.0/spark-2.3.0-bin-hadoop2.7.tgz
```

Extract the `.tgz` zip file (see Figures 2-52 and 2-53) and rename the directory spark-2.3.0:

```
tar - xvf spark-2.3.0-bin-hadoop2.7.tgz
```

```
vagrant@SparkMachine:~$ tar -xvf spark-2.3.0-bin-hadoop2.7.tgz
```

Figure 2-52. *Extract the* `.tgz` *zip file*

```
mv spark-2.3.0-bin-hadoop2.7 spark-2.3.0
```

```
vagrant@SparkMachine:~$ mv spark-2.3.0-bin-hadoop2.7 spark-2.3.0
```

Figure 2-53. *Unzip the Spark binaries*

```
vagrant@SparkMachine:~$ ls
java8   postinstall.sh   spark-2.3.0   spark-2.3.0-bin-hadoop2.7.tgz
```

Next, set the SPARK_HOME in the .bashrc profile file as shown in Figure 2-54).
For example:

```
export SPARK_HOME=<your_spark_installation_path>
export PATH=$PATH:$SPARK_HOME/bin:$SPARK_HOME/sbin
```

```
vagrant@SparkMachine:~$ sudo vi ~/.bashrc
```

Figure 2-54. *Editing the .bashrc file*

Add the following lines to the end of the .bashrc file:

```
export SPARK_HOME=/home/vagrant/spark-2.3.0
export PATH=$PATH:$JAVA_HOME/bin
```

Note   The installation path could be different from /home/vagrant/
spark-2.3.0.

Once the PATH is added, use source ~/.bashrc to update the .bashrc file to the same
session without restarting the machine. Then verify the updated path details, as shown
in Figure 2-55.

```
vagrant@SparkMachine:~$ echo $SPARK_HOME
/home/vagrant/spark-2.3.0
```

Figure 2-55. *Verifying the Spark home path*

After installing Spark and updating the PATH variable, specify the slave details (i.e., the worker node details) for the Spark cluster by following these steps.

1. Navigate to the conf directory in the Spark installation folder:

   cd /home/vagrant/spark-2.3.0/conf

2. Rename the slaves.template file slaves:

   mv slaves.template slaves

3. Edit the slaves file and add SparkMachine at the end of the file (see Figure 2-56):

   vi slaves

```
vagrant@SparkMachine:~$ cd ~/spark-2.3.0/
vagrant@SparkMachine:~/spark-2.3.0$ cd conf/
vagrant@SparkMachine:~/spark-2.3.0/conf$ ls
docker.properties.template  log4j.properties.template    slaves
fairscheduler.xml.template  metrics.properties.template  spark-defaults.conf.template
vagrant@SparkMachine:~/spark-2.3.0/conf$ mv slaves.template slaves
vagrant@SparkMachine:~/spark-2.3.0/conf$ vi slaves
```

Figure 2-56. *Configuration - adding slave host names*

4. Rename the spark-env.sh.template file spark-env.sh:

   mv spark-env.sh.template spark-env.sh

5. Edit the spark-env.sh file and add the JAVA_HOME path to the file.

   vi spark-env.sh

6. Add the following line in the spark-env.sh file.

   export JAVA_HOME=/home/vagrant/java8

Check for the running services (see Figure 2-57) to understand that the Spark cluster is not running.

```
vagrant@SparkMachine:~$ jps
1384 Jps
```

Figure 2-57. *Check the running Java processes*

Now, on the SparkMachine machine, start the processes by calling the script start-all.sh (see Figure 2-58), which starts the master and worker processes on the same machine.

```
vagrant@SparkMachine:~$ start-all.sh
starting org.apache.spark.deploy.master.Master, logging to
pache.spark.deploy.master.Master-1-SparkMachine.out
SparkMachine: Warning: Permanently added 'sparkmachine,192.
SparkMachine: starting org.apache.spark.deploy.worker.Worke
-vagrant-org.apache.spark.deploy.worker.Worker-1-SparkMachi
vagrant@SparkMachine:~$ jps
1525 Master
1592 Worker
```

Figure 2-58. *Start the Spark processes*

Also, the master process can be started using start-master.sh and worker processes can be started using start-slaves.sh separately.

Spark Master UI

When the Spark cluster is running, browse the Spark UI using the following URL to learn about worker nodes attached to the master, running applications, and the cluster resources.

```
http://mastermachine-ip_address:8080/
```

```
Example: http://192.168.163.153:8080/
```

The following information can be found in the Spark Master Web UI (see Figure 2-59).

URL: spark://SparkMaster:7077
REST URL: spark://SparkMaster:6066 (cluster mode)
Alive Workers: 2
Cores in use: 4 Total, 0 Used
Memory in use: 2.0 GB Total, 0.0 B Used
Applications: 0 Running, 0 Completed
Drivers: 0 Running, 0 Completed
Status: ALIVE

← → C   ⓘ 192.168.163.153:8080

 Spark Master at spark://SparkMachine:7077

URL: spark://SparkMachine:7077
REST URL: spark://SparkMachine:6066 *(cluster mode)*
Alive Workers: 1
Cores in use: 2 Total, 0 Used
Memory in use: 1024.0 MB Total, 0.0 B Used
Applications: 0 Running, 0 Completed
Drivers: 0 Running, 0 Completed
Status: ALIVE

Workers (1)

Worker Id	Address
worker-20180521062606-192.168.163.153-40706	192.168.163.153:40706

Figure 2-59. *Spark Web UI*

Also, the Workers, Running Applications, and Completed Application details would be updated in the same UI.

Points to Remember

1. We have created three different virtual machines using Oracle VirtualBox.

2. The storage and network settings are very important for smooth installation and usage.

3. Install Java and verify the Java home path without fail.

4. Copy the Spark binaries and edit the configuration files.

5. Start the Spark processes and verify the master and worker details in the Web UI.

CHAPTER 3

Introduction to Apache Spark and Spark Core

In the previous chapters, the fundamental concepts of Scala programming, pure function, pattern matching, singleton objects, Scala collections, and functional programming features of Scala have been covered.

For this chapter, some prior experience with Scala and Hadoop MapReduce is ideal. A mandatory prerequisite for this chapter is to have read the previous chapters.

In this chapter, we are going to discuss the need for Apache Spark, Spark architecture, and Spark Core. We will be focusing on these topics:

- What is Apache Spark?

- Why Apache Spark?

- Spark vs. Hadoop MapReduce.

- Spark architecture.

- Spark Ecosystem.

- Spark Core.

- Resilient distributed data set (RDD) transformation.

- RDD actions.

- Working with pair RDDs.

- Direct Acylic Graph (DAG) in Apache Spark.

- Persisting RDD.

- Simple Build Tool (SBT).

© Subhashini Chellappan, Dharanitharan Ganesan 2018
S. Chellappan and D. Ganesan, *Practical Apache Spark*, https://doi.org/10.1007/978-1-4842-3652-9_3

What Is Apache Spark?

Apache Spark is an open source, fast, general-purpose, in-memory processing engine for big data processing. Apache Spark was developed in 2009 at the University of California Berkeley's AMP Lab and later open sourced as an Apache project in 2010. Apache Spark is written in Scala and provides high-level application programming interfaces (APIs) in Java, Scala, Python, and R.

Note   Apache Spark 1.x is written in Scala 2.10 and Apache Spark 2.x is written in Scala 2.11.

Why Apache Spark?

Let's discuss the need of Spark.

- Apache Spark provides a unified framework to perform different tasks that would have previously required different engines for processing such as batch, real-time processing.

- Apache Spark provides high-level operators (e.g., map, filter, etc.) to process the data that are not available in Hadoop MapReduce.

- Apache Spark is 100 times faster than Hadoop MapReduce when you run Spark in memory and 10 times faster than Hadoop MapReduce even when you run Spark on disk.

- Apache Spark supports both batch processing and real-time processing.

- Apache Spark provides an interactive shell that you can use for learning and exploring data.

- Apache Spark is not bundled with a storage system. Local file systems, Hadoop Distributed File System (HDFS), Cassandra, S3, and others can be used as storage systems.

Spark vs. Hadoop MapReduce

Table 3-1 provides a comparison between Spark and Hadoop MapReduce.

Table 3-1. *Spark vs. Hadoop MapReduce*

	Apache Spark	**Hadoop MapReduce**
Introduction	Open source big data processing framework. Faster and general-purpose data processing engine. Unified framework to process various tasks such as batch, interactive, iterative processing.	Open source framework to process structured and unstructured data that are stored in the Hadoop Distributed File System (HDFS) Hadoop MapReduce processes data only in batch mode
Speed	100 times faster than Hadoop MapReduce when Spark runs in memory and 10 times faster when Spark runs on disk Spark makes this possible by reducing the number of read/write cycles to disk and storing intermediate data in memory	Hadoop MapReduce reads and writes from disk, which slows down the processing speed of Hadoop MapReduce
Difficulty	Spark provides high-level operators such as map, filter, and so on, which makes the developer's job easy	Developers need to hand code each operation in Hadoop MapReduce
Real-time processing	Spark supports both batch and real-time processing	Hadoop MapReduce supports only batch processing
Interactive mode	Spark provides an interactive shell to learn and explore data	Hadoop MapReduce does not provide an interactive shell

Apache Spark Architecture

Apache Spark has a master–slave architecture with a cluster manager and two daemons, as shown in Figure 3-1. The daemons are master and worker.

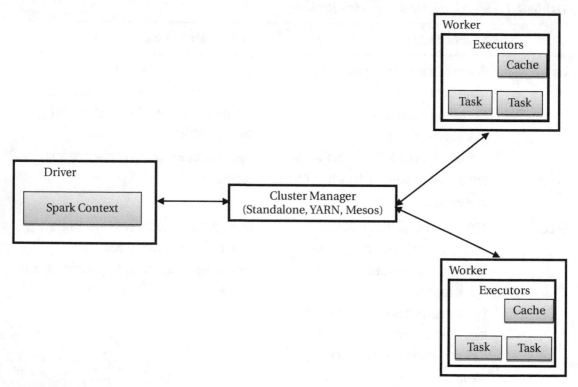

Figure 3-1. *Apache Spark architecture*

At a high level, every Spark application consists of a driver program that is responsible for running the user's main function and executing various parallel operations on the cluster. A Spark cluster consists of a single coordinator called a driver and many distributed workers. The driver communicates in several ways:

- With Spark through the SparkContext object. The SparkContext object is the entry point for Spark functionality. The SparkContext object is available as sc.

- With a large number of distributed workers called executors to execute tasks.

- With Cluster Manager for resource allocation to execute the tasks.

The driver program has the following features.

- It is the place where SparkContext is created.

- It is the entry point for the interactive shell, which is available only for Scala, Python, and R.

- It runs the application's main function.

- It is responsible for scheduling jobs and allocating resources to execute tasks.

- It is responsible for converting user applications into smaller execution units known as tasks.

The executor has several functions.

- It is responsible for executing tasks and performing data processing.

- It reads from and writes to external sources.

- It stores intermediate results in memory.

The Cluster Manager has the following attributes.

- It is an external service.

- It is responsible for acquiring resources such as CPU, memory, and more, and allocating them to Spark applications.

- There are three types of Cluster Manager: stand-alone, yet another resource negotiator (YARN; specific to Hadoop), and Mesos (a general-purpose cluster manager). This is illustrated in Figure 3-2.

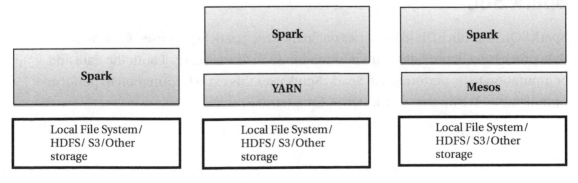

Figure 3-2. *Spark Cluster Manager*

Spark Components

Let's look at the various components of Apache Spark, illustrated in Figure 3-3.

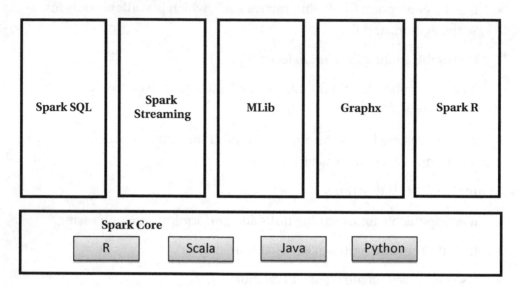

Figure 3-3. *Spark components*

Spark Core (RDD)

Spark Core is known as resilient distributed data set (RDD). This is the fundamental data structure of Spark. RDD is an immutable distributed collection of objects that can be operated on in parallel. This component is used for batch processing. We discuss Spark Core further later in this chapter.

Spark SQL

Spark SQL is to run SQL-like queries on Spark data. Spark SQL is used for structure data processing. It provides extra information such as structure of both the data and computation to be performed to Spark. Spark uses this extra information to perform optimization. We discuss Spark SQL in Chapter 4.

Spark Streaming

Spark Streaming is an extension of the Spark core. Spark Streaming is a scalable, fault-tolerant, high-throughput streaming engine to process live data streams. The Spark streaming component is used for real-time processing. Data can be taken from many sourses such as Kafka, Kinesis, HDFS, Flume, and so on, and can be processed using high-level Spark core APIs such as map, filter, join, and so on. In Chapter 5, we discuss Spark Streaming in detail.

MLib

Spark MLib is a machine learning library for Spark. Spark MLib is for scalable practical machine learning. Spark MLib is focused on common learning algorithms such as regression, classification, clustering, and colloborative filtering. In Chapter 8, we discuss Spark MLib.

GraphX

GraphX is for graphs and grpah parallel computation. Spark GraphX extends RDD by introducing a new Graph abstraction, a directed multigraph with properties attached to each vertex and edge.

SparkR

SparkR is a package for R that provides a lightweight front end to use Apache Spark from R. In Spark 2.3.0, we have data frame distribution, which is similar to the R data frame to perform selection, filtering, aggregation, and so on, on large data sets. SparkR also supports distributed machine learning using MLib. In Chapter 9, we cover the SparkR component further.

Spark Shell

Spark provides an interactive shell for data exploration and testing, a read, evaluate, print loop (REPL). To start Spark Shell, type `spark-shell` at the command line (see Figure 3-4). Refer to Chapter 2 for Spark installation and cluster setup.

```
Spark context available as 'sc' (master = local[*], app id = local-1517754512031).
Spark session available as 'spark'.
Welcome to


    ____              __
   / __/__  ___ _____/ /__
  _\ \/ _ \/ _ `/ __/  '_/
 /___/ .__/\_,_/_/ /_/\_\   version 2.1.0
    /_/

Using Scala version 2.11.8 (Java HotSpot(TM) 64-Bit Server VM, Java 1.8.0_77)
Type in expressions to have them evaluated.
Type :help for more information.

scala> sc.appName
res0: String = Spark shell
```

Figure 3-4. *Starting Spark Shell*

Note   We are going to use Spark Shell to work with RDD. In Spark Shell, SparkContext is available as sc and can be used to perform various data processing tasks.

Spark Core: RDD

The fundamental data structure of Spark is RDD, a fault-tolerant collection of elements that can be operated on in parallel. It is resilient because it has built-in fault tolerance. If something goes wrong, we can reconstruct it from the source (lineage). We discuss this later in more detail. Data are distributed in memory across the worker nodes. A data set represents records of the data.

RDD is an immutable collection of distributed objects, elements partitioned across the nodes of clusters and operated on in parallel as shown in Figure 3-5.

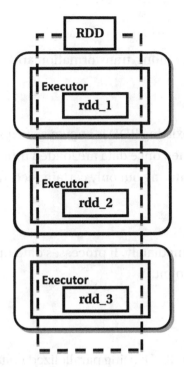

Figure 3-5. *RDD: Resilient distributed data set*

RDD has the following characteristics.

- *In-memory computation:* RDD stores intermediate results in distributed memory.

- *Lazy evaluations:* In Spark all transformations are lazy. Lazy means they do not compute their results until it is required.

- *Fault tolerance:* RDD rebuilds the lost data automatically from the source on failure using lineage. Each RDD remembers how it was created from other data sets.

- *Immutability:* RDDs are immutable in the sense thaat data cannot be modified in place. RDDs can be modified only by applying RDD operations, namely transformation and action.

- *Partitioning:* Data are divided into partitions and distributed across the cluster and operated in parallel.

- *Action/transformations:* In Spark RDD, all the computations are either actions or transformations.

RDD Operations

RDD provides two types of operations, transformation and actions.

Transformations

Creating a new RDD from an existing RDD is known as *transformation.* Chain of transformation can be performed once data are loaded into memory. An example is extracting specific fields and filtering out only certain records.

Actions

Spark doesn't process data immediately. It processes data only when an action is called. Examples include a sum or a count.

Creating an RDD

There are two ways to create an RDD: using parallelized collection or from an external data source. Let's look at each of these methods next.

Using Parallelized Collection

Parallelized collections can be created by calling SparkContext's parallelize method on an existing collection. When the parallelize method is applied on a collection, elements of the collection are copied to form a distributed data set.

Objective: To create an RDD.

Action: Use the parallelize method of SparkContext. Create Array of integers and pass that as an argument to the parallelize method.

```
val rdd = Array(1, 2, 3, 4, 5)        //Line 1

val rdd1 = sc.parallelize(rdd)        //Line 2

rdd1.collect()                        //Line 3
```

Output: Use the collect() action, which returns all the elements of the data set as an array to the driver program, to display the output of the RDD displayed in Figure 3-6.

```
scala> rdd1.collect()
res1: Array[Int] = Array(1, 2, 3, 4, 5)
```

Figure 3-6. *RDD output*

From External Data Source

Spark can create distributed data sets from any storage system such as Hadoop, Cassandra, and so on.

Objective: To create an RDD using an external data source.
Input File: keywords.txt (see Figure 3-7).

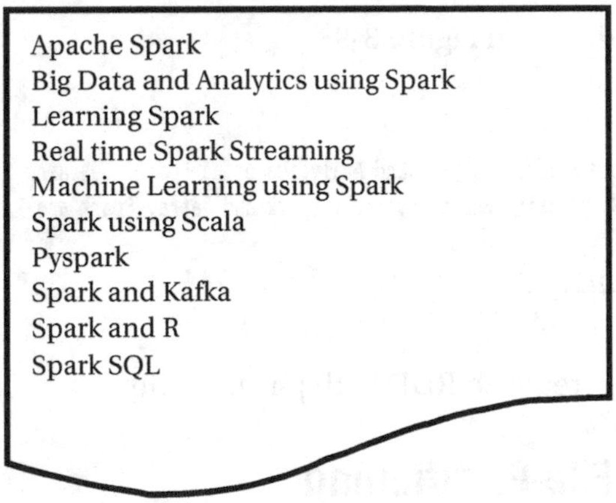

Apache Spark
Big Data and Analytics using Spark
Learning Spark
Real time Spark Streaming
Machine Learning using Spark
Spark using Scala
Pyspark
Spark and Kafka
Spark and R
Spark SQL

Figure 3-7. *Keywords.txt file*

Action: Use the textFile() method of SparkContext. Specify the URL path of the local file system as an argument to the textFile() method.

```
val rdd = sc.textFile("/home/data/keywords.txt")   //Line 1
```

Output: The RDD output is shown in Figure 3-8.

```
scala> rdd.collect
res2: Array[String] = Array(Apache Spark, Big Data and Analytics using Spark, Learning Spark, Real time Spark Streaming,
achine Learning using Spark, Spark using Scala, Pyspark, Spark and Kafka, Spark and R, Spark SQL)

    -
```

Figure 3-8. *RDD output*

Note   Store the file in the local file system.

Let's discuss how to create an RDD from HDFS.

Creating an RDD from the Hadoop File System

Objective: To create an RDD using external data source - HDFS.
Input File: keywords.txt (as shown earlier in Figure 3-7).
Action: Use textFile() of SparkContext. Specify the URL path of the HDFS as an argument to the textFile() method.

```
val rdd = sc.textFile("hdfs://localhost:9000/data/keywords.txt") //Line 1
```

Output: RDD output is shown in Figure 3-9.

```
scala> rdd.collect
res2: Array[String] = Array(Apache Spark, Big Data and Analytics using Spark, Learning Spark, Real time Spark Streaming
 Machine Learning using Spark, Spark using Scala, Pyspark, Spark and Kafka, Spark and R, Spark SQL)
```

Figure 3-9. *RDD output*

Let's discuss next how to create an RDD with partitioning.

Creating an RDD: File Partitioning

Spark divides data into partitions and distributes them across a cluster. By default, it divides data into two partitions, but the number of partitions can be specified while creating an RDD as shown here.

```
textFile(filename, minPartitions)
```

```
val rdd = sc.textFile("home/data/keywords.txt", 3)
```

Here, the number of partitions is three, so the file will be divided into three partitions.

RDD Transformations

Let's discuss the various transformations provided by Apache Spark.

1. map(func): Returns a new data set by operating on each element of the source RDD.

Objective: To illustrate a map(func) transformation.

Action: Create an RDD of a numeric list. Then apply map(func) to multiply each element by 2.

```
val mapRdd = sc.parallelize(List(1, 2, 3, 4))          // Line 1

val mapRdd1 = mapRdd.map(x => x * 2)                    // Line 2
```

Output: The mapRdd1 data set is shown in Figure 3-10.

```
scala> mapRdd1.collect
res1: Array[Int] = Array(2, 4, 6, 8)
```

***Figure 3-10.** mapRdd1 data set*

2. flatMap(func): Like map, but each item can be mapped to zero, one, or more items.

Objective: To illustrate the flatMap(func) tranformation.

Action: Create an RDD for a list of Strings, apply flatMap(func).

```
val flatMapRdd = sc.parallelize(List("hello world", "hi"))    //Line 1

val flatMapRdd1= flatMapRdd.flatMap(line => line.split(" ")) //Line 2
```

Output: The flatMapRdd data set is shown in Figure 3-11.

```
scala> flatMapRdd1.collect
res2: Array[String] = Array(hello, world, hi)
```

***Figure 3-11.** flatMapRdd1 data set*

Apply map(func) in line 2 instead of flatMap(func).

```
val mapRdd1= flatMapRdd.map(line => line.split(" "))              //Line 2
```

Output: The mapRdd1 output is shown in Figure 3-12.

```
scala> mapRdd1.collect
res3: Array[Array[String]] = Array(Array(hello, world), Array(hi))
```

Figure 3-12. mapRdd1 data set

3.   filter(func): Returns a new RDD that contains only elements
     that satisfy the condition.

Objective: To illustrate filter(func) tranformation.
Action: Create an RDD using an external data set. Apply filter(func) to display the lines that contain the word Kafka.
Input File: keywords.txt (refer to Figure 3-7).

```
val filterRdd = sc.textFile("/home/data/keywords.txt")     //Line 1
```

```
val filterRdd1 = filterRdd.filter(line => line.contains("Kafka"))//Line 2
```

Output: The filterRdd1 data set is shown in Figure 3-13.

```
scala> filterRdd1.collect
res5: Array[String] = Array(Spark and Kafka)
```

Figure 3-13. filterRdd1 data set

4.   mapPartitions(func): It is similar to map, but works on the
     partition level.

Objective: To illustrate the mapPartitions(func) tranformation.
Action: Create an RDD of numeric type. Apply mapPartition(func)

```
val rdd = sc.parallelize(10 to 90)                          //Line 1
```

```
rdd.mapPartitions( x => List(x.next).iterator).collect //Line 2
```

Output: The output is shown in Figure 3-14.

```
scala> rdd.mapPartitions( x => List(x.next).iterator).collect
res0: Array[Int] = Array(10, 50)
```

Figure 3-14. *mapPartition output*

Here, the data set is divided into two partitions. Partition1 contains elements 10 to 40 and partition2 contains elements 50 to 90.

5. mapPartitionsWithIndex(func): This is similar to mapPartitions,but provides a function with an Int value to indicate the index position of the partition.

Objective: To illustrate the mapPartitionsWithIndex(func) tranformation.

Action: Create an RDD of numeric type. Apply mapPartitionWithIndex(func) to display the position of each element in the partition.

```
val rdd = sc.parallelize(1 to 5, 2)  // Line 1
```

```
rdd.mapPartitionsWithIndex( (index: Int, it: Iterator[Int]) => it.toList.
map(x => index + ", "+x).iterator).collect  //Line 2
```

Output: The output is shown in Figure 3-15.

```
res0: Array[String] = Array(0, 1, 0, 2, 1, 3, 1, 4, 1, 5)
```

Figure 3-15. *mapPartitionsWithIndex output*

Here, partition1 contains elements 1 and 2, whereas partition2 contains elements 3, 4, and 5.

6. union(otherDataset): This returns a new data set that contains the elements of the source RDD and the argument RDD. The key rule here is the two RDDs should be of the same data type.

Objective: To illustrate union(otherDataset) .

Action: Create two RDDs of numeric type as shown here. Apply union(otherDataset) to combine both RDDs.

```
val rdd = sc.parallelize(1 to 5)                 //Line 1
```

```
val rdd1 = sc.parallelize(6 to 10)               //Line 2
```

```
val unionRdd=rdd.union(rdd1)                      //Line 3
```

Output: The unionRdd data set is shown in Figure 3-17.

```scala
scala> unionRdd.collect
res7: Array[Int] = Array(1, 2, 3, 4, 5, 6, 7, 8, 9, 10)
```

Figure 3-16. *unionRdd data set*

7. intersection(otherDataset): This returns a new data set that
   contains the intersection of elements from the source RDD and
   the argument RDD.

Objective: To illustrate intersection(otherDataset).
Action: Create two RDDs of numeric type as shown here. Apply
intersection(otherDataset) to display all the elements of source RDD that also belong
to argument RDD.

```scala
val rdd = sc.parallelize(1 to 5)                    //Line 1

val rdd1 = sc.parallelize(1 to 2)                   //Line 2

val intersectionRdd = rdd.intersection(rdd1)        //Line 3
```

Output: The intersectionRdd data set is shown in Figure 3-17.

```scala
scala> intersectionRdd.collect
res8: Array[Int] = Array(2, 1)
```

Figure 3-17. *The intersectionRdd data set*

8. distinct([numTasks]): This returns a new RDD that contains
   distinct elements within a source RDD.

Objective: To illustrate distinct([numTasks]).
Action: Create two RDDs of numeric type as shown here. Apply union(otherDataset)
and distinct([numTasks]) to display distinct values.

```scala
val rdd = sc.parallelize(10 to 15)            //Line 1

val rdd1 = sc.parallelize(10 to 15)           //Line 2

val distinctRdd=rdd.union(rdd1).distinct      //Line 3
```

Output: The `distinctRdd` data set is shown in Figure 3-18.

```
scala> distinctRdd.collect
res9: Array[Int] = Array(12, 13, 14, 10, 15, 11)
```

Figure 3-18. *distinctRdd data set*

RDD Actions

Action returns values to the driver program. Here we discuss RDD actions.

1. `reduce(func)`: This returns a data set by aggregating the elements of the data set using a function `func`. The function takes two arguments and returns a single argument. The function should be commutative and associative so that it can be operated in parallel.

Objective: To illustrate `reduce(func)`.
Action: Create an RDD that contains numeric values. Apply reduce(func) to display the sum of values.

```
val rdd = sc.parallelize(1 to 5)                //Line 1

val sumRdd = rdd.reduce((t1,t2) => t1 + t2)     //Line 2
```

Output: The sumRdd value is shown in Figure 3-19.

```
scala> val sumRdd = rdd.reduce((t1,t2) => t1 + t2)
sumRdd: Int = 15
```

Figure 3-19. *sumRdd value*

2. `collect()`: All the elements of the data set are returned as an array to the driver program.

Objective: To illustrate `collect()`.
Action: Create an RDD that contains a list of strings. Apply `collect` to display all the elements of the RDD.

```
val rdd = sc.parallelize(List("Hello Spark", "Spark Programming")) //Line 1

rdd.collect()                                                      //Line 2
```

Output: The result data set is shown in Figure 3-20.

```
scala> rdd.collect
res10: Array[String] = Array(Hello Spark, Spark Programming)
```

Figure 3-20. *The result data set*

    3.   count(): This returns the number of elements in the data set.

Objective: To illustrate count().

Action: Create an RDD that contains a list of strings. Apply count to display the number of elements in the RDD.

```
val rdd = sc.parallelize(List("Hello Spark", "Spark Programming")) //Line 1

rdd.count()                                                          //Line 2
```

Output: The number of elements in the data set is shown in Figure 3-21.

```
scala> rdd.count
res11: Long = 2
```

Figure 3-21. *The number of elements in the data set*

    4.   first(): This returns the first element in the data set.

Objective: To illustrate first().

Action: Create an RDD that contains a list of strings. Apply first() to display the first element in the RDD.

```
val rdd = sc.parallelize(List("Hello Spark", "Spark Programming")) //Line 1

rdd.first()                                                         //Line 2
```

Output: The first element in the data set is shown in Figure 3-22.

```
scala> rdd.first()
res12: String = Hello Spark
```

Figure 3-22. *First element in the data set*

5. take(n): This returns the first n elements in the data set as an array.

Objective: To illustrate take(n) .

Action: Create an RDD that contains a list of strings. Apply take(n) to display the first n elements in the RDD.

```
val rdd = sc.parallelize(List("Hello","Spark","Spark SQL","MLib")) //Line 1

rdd.take(2)                                              //Line 2
```

Output: The first n elements in the data set are shown in Figure 3-23.

```
scala> rdd.take(2)
res14: Array[String] = Array(Hello, Spark)
```

Figure 3-23. *First n elements in the data set*

6. saveAsTextFile(path): Write the elements of the RDD as a text file in the local file system, HDFS, or another storage system.

Objective: To illustrate saveAsTextFile(path).

Action: Create an RDD that contains a list of strings. Apply saveAsTextFile(path) to write the elements in the RDD to a file.

```
val rdd = sc.parallelize(List("Hello","Spark","Spark SQL","MLib")) //Line 1

rdd. saveAsTextFile("/home/data/output")                 //Line 2
```

7. foreach(func): foreach(func) operates on each element in the RDD.

Objective: To illustrate foreach(func) .

Action: Create an RDD that contains a list of strings. Apply foreach(func) to print each element in the RDD.

```
val rdd = sc.parallelize(List("Hello","Spark","Spark SQL","MLib")) //Line 1

rdd.foreach(println)                                     //Line 2
```

Output: The output of foreach(println) is shown in Figure 3-24.

```
scala> rdd.foreach(println)
Hello
Spark
Spark SQL
MLib
```

Figure 3-24. *The foreach(println) output*

Working with Pair RDDs

Pair RDDs are special form of RDD. Each element in the pair RDDs is represented as a key/value pair. Pair RDDs are useful for sorting, grouping, and other functions. Here we introduce a few pair RDD transformations.

1. groupByKey([numTasks]): When we apply this on a data set of
   (K, V) pairs, it returns a data set of (K, Iterable<V>) pairs.

Objective: To illustrate the groupByKey([numTasks]) transformation. Display names by each country.

Input File: people.csv (see Figure 3-25).

```
year,name,country,count
2015,john,us,215
2016,jack,ind,120
2017,james,ind,56
2018,john,cannada,67
2016,james,us,218
```

Figure 3-25. *People.csv file*

Action: Follow these steps.
1. Create an RDD using people.csv.
2. Use the filter(func) transformation to remove the header line.
3. Use map(func) and the split() method to split fields by ",".
4. Retrieve country, name field using map(func).
5. Apply groupByKey to group names by each country.

Note   The field index starts from 0.

```
val rdd = sc.textFile("/home/data/people.csv")              //Line 1

val splitRdd = rdd.filter(line => !line.contains
("year")).map(line => line.split(","))                      //Line 2

val  fieldRdd= splitRdd.map(f => (f(2),f(1)))               //Line 3

val groupNamesByCountry=fieldRdd.groupByKey                 //Line 4

groupNamesByCountry.foreach(println)                        //Line 5
```

Output: The data set that contains names by each country is shown in Figure 3-26.

```
scala> groupNamesByCountry.foreach(println)
(ind,CompactBuffer(jack, james))
(us,CompactBuffer(john, james))
(cannada,CompactBuffer(john))
```

Figure 3-26. *RDD that contains names by country*

2. reduceByKey (func, [numTasks]): When reduceByKey(func, [numTasks]) is applied on a data set of (K, V) pairs, it returns a data set of (K, V) pairs. Here the values for each key are aggregated using reduce(func). The func should be of type (V,V) => V.

Objective: To illustrate the reduceByKey (func, [numTasks]) transformation. Display the total names count by each name.

Input File: people.csv (refer Figure 3-25).

Action:

1. Create an RDD using people.csv.

2. Use the filter(func) transformation to remove the header line.

3. Use map(func) and split () to split fields by ",".

4. Retrieve name, count field using map(func).

5. Apply reduceByKey(func) to count names.

```
val rdd = sc.textFile("/home/data/people.csv")                    //Line 1

val splitRdd = rdd.filter(line => !line.contains
("year")).map(line => line.split(","))                            //Line 2

val fieldRdd = splitRdd.map(f => (f(1),f(3).toInt))               //Line 3

val namesCount=fieldRdd.reduceByKey((v1,v2) => v1 + v2)           //Line 4

namesCount.foreach(println)                                       //Line 5
```

Output: The data set that contains name counts by each name is shown in Figure 3-27.

```
scala> namesCount.foreach(println)
(james,274)
(jack,120)
(john,282)
```

Figure 3-27.   *RDD that contains name counts by each name*

3. sortByKey([ascending], [numTasks]): When
    sortByKey([ascending], [numTasks]) is applied on a data set
    of (K, V) pairs, it returns a data set of (K, V) pairs where keys
    are sorted in ascending or descending order as specified in the
    boolean ascending argument.

Objective: To illustrate the sortByKey([ascending], [numTasks]) transformation.
Display the names in ascending order.

Input File: people.csv (refer to Figure 3-25).

Action:

1. Create an RDD using people.csv.

2. Use the filter(func) transformation to remove the header line.

3. Use map(func) and split() to split fields by ",".

4. Retrieve name, count field using map(func).

5. Apply sortByKey(func) to display names in ascending order.

```
val rdd = sc.textFile("/home/data/people.csv")                    //Line 1

val splitRdd = rdd.filter(line => !line.contains
("year")).map(line => line.split(","))                            //Line 2

val fieldRdd = splitRdd.map(f => (f(1),f(3).toInt)).sortByKey() //Line 3

fieldRdd.foreach(println)                                         //Line 4
```

Output: The data set that contains names in ascending order is shown in Figure 3-28.

```
scala> fieldRdd.foreach(println)
(jack,120)
(james,56)
(james,218)
(john,215)
(john,67)
```

Figure 3-28. *RDD that contains names in ascending order*

Direct Acylic Graph in Apache Spark

DAG in Apache Spark is a set of vertices and edges. In Spark, vertices represent RDDs and edges represent the operation to be applied on RDD. Each edge in the DAG is directed from one vertex to another. Spark creates the DAG when an action is called.

How DAG Works in Spark

At a high level, when an action is called on the RDD, Spark creates the DAG and submits the DAG to the DAG scheduler.

1. The DAG scheduler divides operators such as map, flatMap, and so on, into stages of tasks.

2. The result of a DAG scheduler is a set of stages.

3. The stages are passed on to the Task Scheduler.

4. The Task Scheduler launches tasks via Cluster Manager.

5. The worker executes the tasks.

Note   A stage is comprised of tasks based on partitions of the input data.

At a high level, Spark applies two transformations to create a DAG. The two transformations are as follows:

- *Narrow transformation:* The operators that don't require the data to be shuffled across the partitions are grouped together as a stage. Examples are map, filter, and so on.

- *Wide transformation:* The operators that require the data to be shuffled are grouped together as a stage. An example is `reduceByKey`.

DAG visualization can be viewed through the Web UI (`http:/localhost:4040/jobs/`). Scala code to count the occurrence of each word in a file is shown here.

```
sc.textFile("/home//keywords.txt").flatMap(line => line.split(" ")).
map(word => (word,1)).reduceByKey(_+_).collect()
```

Refer to Figure 3-29 for the DAG visualization of word count. The word count problem consists of two stages. The operators that do not require shuffling (`flatMap()` and `map()` in this case) are grouped together as Stage 1 and the operators that require shuffling (`reduceByKey`) are grouped together as Stage 2.

▼ DAG Visualization

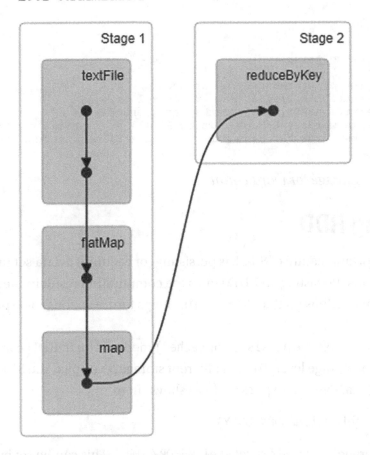

Figure 3-29. *The DAG visualization for word count*

How Spark Achieves Fault Tolerance Through DAG

Spark maintains each RDD's lineage (i.e., previous RDD on which it depends) that is created in DAG to achieve fault tolerance. When any node crashes, Spark Cluster Manager assigns another node to continue processing. Spark does this by reconstructing the series of operations that it should compute on that partition from the source.

To view the lineage, use toDebugString. A lineage graph for word count is shown in Figure 3-30.

```
scala> wc.toDebugString
res1: String =
(2) ShuffledRDD[4] at reduceByKey at <console>:24 []
 +-(2) MapPartitionsRDD[3] at map at <console>:24 []
    |  MapPartitionsRDD[2] at flatMap at <console>:24 []
    |  /home/keywords.txt MapPartitionsRDD[1] at textFile at <console>:24 []
    |  /home/keywords.txt HadoopRDD[0] at textFile at <console>:24 []
```

Figure 3-30. *Lineage for word count*

Persisting RDD

The most important feature of Spark is persisting (or caching) a data set in memory across operations. Persisting an RDD stores the computation result in memory and reuses it in other actions on that data set. This helps future actions to be performed much faster.

To persist an RDD, use the persist() or cache() methods on it. RDD can be persisted using a different storage level. To set a different storage level, pass the StorageLevel object (Scala, Java, Python) to persist() as shown here.

persist(StorageLevel.MEMORY_ONLY)

The default storage level is StorageLevel.MEMORY_ONLY. This can be set by using the cache() method.

Spark persists shuffle operations (e.g., reduceByKey) with intermediate data automatically even without calling the persist method. This avoids recomputation of the entire input if a node fails during the shuffle. Table 3-2 shows different storage levels.

Table 3-2. *Storage Level*

Storage Level	Meaning
MEMORY_ONLY	Store RDD as deserialized Java objects in the Java Virtual Machine. If the RDD does not fit in memory, some partitions will not be cached and will be recomputed on the fly each time they're needed. This is the default level.
MEMORY_AND_DISK	Store RDD as deserialized Java objects in the Java Virtual Machine. If the RDD does not fit in memory, store the partitions that don't fit on disk, and read them from there when they're needed.
MEMORY_ONLY_SER (Java and Scala)	Store RDD as serialized Java objects (one byte array per partition). This is generally more space-efficient than deserialized objects, especially when using a fast serializer, but more CPU-intensive to read.
MEMORY_AND_DISK_SER (Java and Scala)	Similar to MEMORY_ONLY_SER, but spill partitions that don't fit in memory to disk instead of recomputing them on the fly each time they're needed.
DISK_ONLY	Store the RDD partitions only on disk.
MEMORY_ONLY_2, MEMORY_AND_DISK_2, etc.	Same as the levels above, but replicate each partition on two cluster nodes.

Shared Variables

Normally, Spark executes RDD operations such as map or reduce on a remote cluster node. When a function is passed to RDD operation, it works on the separate copies of all the variables used in the function. All these variables are copied to each machine and updates to the variables are not propagated back to the driver. This makes read-write across tasks inefficient. To resolve this, Spark provides two common types of shared variables, namely broadcast variables and accumulators. We discuss broadcast variables first.

Broadcast Variables

Broadcast variables help to cache a read-only variable on each machine rather than shipping a copy of it with tasks. Broadcast variables are useful to give a copy of large data set to every node in an efficient manner.

Broadcast variables are created by calling (v) as shown here.

```
val broadcastVar = sc.broadcast(Array(1, 2, 3)) //Line 1
```

Broadcast variables can be accessed by calling the value method as shown in Figure 3-31.

```
broadcastVar.value                                    //Line 2
```

```
scala> broadcastVar.value
res0: Array[Int] = Array(1, 2, 3)
```

Figure 3-31. *Broadcast variable* value *output*

Note   Do not modify object v after it is created to ensure that all nodes get the same value of the broadcast variable.

Accumulators

Accumulators are variables that can be used to aggregate variables across the executors. Accumulators can be used to implement counters or sums. Spark supports accumulators of numeric type by default and programmers can add support for new types.

A numeric accumulator can be created by calling SparkContext.longAccumulator() to accumulate the value of Long. Tasks can add value to the accumulator by using the add method. However, tasks cannot read the accumulator value. Only the driver program can read the accumulator's value.

The following code accumulates the value of an Array.

```
val accum = sc.longAccumulator("My Counter")                    //Line 1

sc.parallelize(Array(10,20,30,40)).foreach(x => accum.add(x))   //Line 2

accum.value                                                     //Line 3
```

The accumulator value can be accessed by calling the `value` method as shown in Figure 3-32.

```
scala> accum.value
res4: Long = 100
```

Figure 3-32.   *Accumulator output*

The accumulator value can be accessed through the Web UI as well (see Figure 3-33).

▾ Aggregated Metrics by Executor							
Executor ID ▴	Address		Task Time	Total Tasks	Failed Tasks	Killed Tasks	Succeeded Tasks
driver	192.168.163.151:46516		97 ms	2	0	0	2

Accumulators	
Accumulable	Value
My Counter	100

Tasks (2)

Index ▴	ID	Attempt	Status	Locality Level	Executor ID / Host	Launch Time	Duration	GC Time	Accumulators	Errors
0	2	0	SUCCESS	PROCESS_LOCAL	driver / localhost	2018/02/12 05:06:03	1 ms		My Counter: 30	
1	3	0	SUCCESS	PROCESS_LOCAL	driver / localhost	2018/02/12 05:06:03	2 ms		My Counter: 70	

Figure 3-33.   *Accumulator value display in Web UI*

Simple Build Tool (SBT)

SBT is a Scala-based build tool for Scala applications. We discuss how to build Spark applications using SBT and submit them to the Spark Cluster.

You can download the latest version of SBT from `http://www.scala-sbt.org/download.html`. Click on the installer and follow the instruction to install SBT.

Let's discuss how we can build a Spark application using SBT.

1.   Create a folder structure as shown in Figure 3-34.

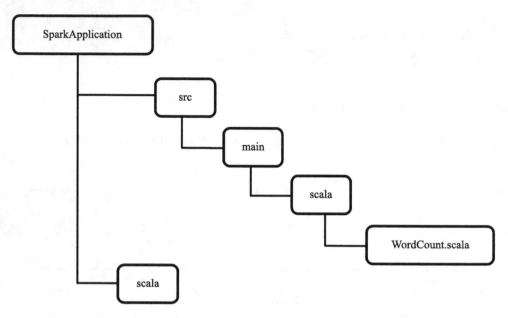

Figure 3-34. *Spark application folder structure*

2.   Create build.sbt as shown in Figure 3-35. Specify all the required libraries.

```
name := "spark-wc"
version := "1.0"
scalaVersion := "2.11.8"
libraryDependencies += "org.apache.spark" % "spark-core_2.11" % "2.1.0"
```

Figure 3-35. *build.sbt file*

3. Write a Spark application to count the occurrence of each word in the keywords.txt file. Consider Figure 3-7 for the input file. The Scala code is shown here.

```scala
package com. Book

import org.apache.spark.{SparkContext, SparkConf}

object WordCount {
    def main(args: Array[String]) {
val conf = new SparkConf().setAppName("Spark WordCount Application")
val sc = new SparkContext(conf)

val inputFileName = args(0)
val outputFileName = args(1)

sc.textFile(inputFileName)
  .flatMap(line => line.split(" "))
  .map(word => (word,1))
  .reduceByKey(_ + _)
  .saveAsTextFile(outputFileName)
  }
        }
```

4. Open a command prompt and navigate to the folder where the Spark word count application is present. Type sbt clean package to build the project.

5. Project, the target directory, will be created as shown in Figure 3-36.

Name	Date modified	Type	Size
project	12-02-2018 12:25	File folder	
src	11-01-2018 12:50	File folder	
target	12-02-2018 12:27	File folder	
build	07-12-2017 12:32	SBT File	1 KB

Figure 3-36. *Project, the target directory*

6. The executable jar will be created inside the target directory,
    `scala-2.11` as shown in Figure 3-37.

Name	Date modified	Type	Size
classes	12-02-2018 12:27	File folder	
resolution-cache	12-02-2018 12:27	File folder	
spark-wc_2.11-1.0	12-02-2018 12:27	Executable Jar File	5 KB

Figure 3-37. *Spark word count jar*

7. Copy the executable jar (`spark-wc_2.11-1.0.jar`) to the Spark
    cluster as shown in Figure 3-38.

Name	Size	Changed	Rights	Owner
..		01-02-2018 10:12:07	rwxr-xr-x	vagrant
keywords.txt	1 KB	01-02-2018 10:12:30	rw-rw-r--	vagrant
people.csv	1 KB	05-02-2018 13:13:43	rw-rw-r--	vagrant
spark-wc_2.11-1.0.jar	5 KB	24-01-2018 10:18:11	rw-rw-r--	vagrant

Figure 3-38. *Spark word count jar*

8. Issue the `spark-submit` command as shown here.

```
spark-submit --class com.book.WordCount --master spark://
masterhostname:7077 /home/data/spark-wc_2.11-1.0.jar /home/data/keywords.
txt /home/data/output
```

Note   Here, the Spark stand-alone Cluster Manager (`spark://`
`masterhostname:7077`) is used to submit the job.

---

9. Output will be created as part of a file as shown in Figures 3-39
    and 3-40.

Name	Size	Changed	Rights	Owner
⬆ ..		01-02-2018 10:12:07	rwxr-xr-x	vagrant
output		12-02-2018 12:58:35	rwxrwxr-x	vagrant
keywords.txt	1 KB	01-02-2018 10:12:30	rw-rw-r--	vagrant
people.csv	1 KB	05-02-2018 13:13:43	rw-rw-r--	vagrant
spark-wc_2.11-1.0.jar	5 KB	12-02-2018 12:45:48	rw-rw-r--	vagrant

Figure 3-39. *Output directory*

Figure 3-40 shows part of the file inside the output directory.

Name	Size	Changed	Rights	Owner
⬆ ...		12-02-2018 12:58:01	rwxrwxr-x	vagrant
_SUCCESS	0 KB	12-02-2018 12:58:35	rw-r--r--	vagrant
part-00000	1 KB	12-02-2018 12:58:34	rw-r--r--	vagrant
part-00001	1 KB	12-02-2018 12:58:34	rw-r--r--	vagrant

Figure 3-40. *Part of a file inside the output directory*

10. Open the part-00000 file to check the output as shown in
     Figures 3-41 and 3-42.

```
(Kafka,1)
(Real,1)
(R,1)
(Big,1)
(Pyspark,1)
(Apache,1)
(SQL,1)
(Analytics,1)
(using,3)
(Scala,1)
(Data,1)
(Streaming,1)
(Learning,2)
```

Figure 3-41. *Word count output*

111

```
(Spark,9)
(Machine,1)
(time,1)
(and,3)
```

Figure 3-42. *Word count output*

Assignments

1. Consider the the sample `logs.txt` shown in Figure 3-43. Write a Spark application to count the total number of WARN lines in the `logs.txt` file.

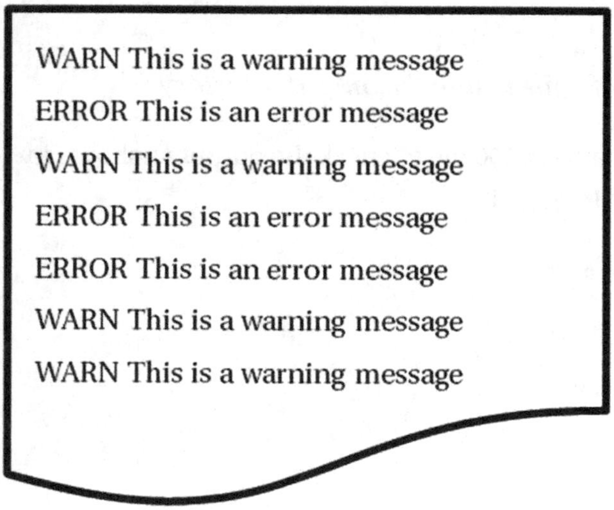

WARN This is a warning message

ERROR This is an error message

WARN This is a warning message

ERROR This is an error message

ERROR This is an error message

WARN This is a warning message

WARN This is a warning message

Figure 3-43. *Sample* `logs.txt`

Reference Links

1. https://spark.apache.org/docs/latest/rdd-programming-guide.html

Points to Remember

- Apache Spark is 100 times faster than Hadoop MapReduce.

- Apache Spark has a built-in real-time stream processing engine to process real-time data.

- RDD is an immutable collection of objects.

- RDD supports two types of operations: transformation and actions.

- Pair RDDs are useful to work with key/value data sets.

- Broadcast and accumulators are shared variables.

- SBT can be used to build Spark applications.

In the next chapter, we discuss how to deal with structure data using Spark SQL.

CHAPTER 4

Spark SQL, DataFrames, and Datasets

In the previous chapter on Spark Core, you learned about the RDD transformations and actions as the fundamentals and building blocks of Apache Spark. In this chapter, you will learn about the concepts of Spark SQL, DataFrames, and Datasets. As a heads up, the Spark SQL DataFrames and Datasets APIs are useful to process structured file data without the use of core RDD transformations and actions. This allows programmers and developers to analyze the structured data much faster than they would by applying the transformations on RDDs created.

The recommended background for this chapter is to have some prior experience with Java or Scala. Experience with any other programming language is also sufficient. Also, having some familiarity with the command line is beneficial.

The mandatory prerequisite for this chapter is to have completed the previous chapter on Spark Core, practiced all the demos, and completed all the hands-on exercises given in the previous chapter.

By end of this chapter, you will be able to do the following:

- Understand the concepts of Spark SQL.

- Use the DataFrames and Datasets APIs to process the structured data.

- Run traditional SQL queries on structured file data.

Note   It is recommended that you practice the code snippets provided as the illustrations and practice the exercises to develop effective knowledge of Spark SQL concepts and DataFrames, and the Datasets API.

© Subhashini Chellappan, Dharanitharan Ganesan 2018
S. Chellappan and D. Ganesan, *Practical Apache Spark*, https://doi.org/10.1007/978-1-4842-3652-9_4

What Is Spark SQL?

Spark SQL is the Spark module for processing structured data. The basic Spark RDD APIs are used to process semistructured and structured data with the help of built-in transformations and actions. The Spark SQL APIs, though, help developers to process structured data without applying transformations and actions. The DataFrame and Datasets APIs provide several ways to interact with Spark SQL.

Datasets and DataFrames

Dataset is a new interface added in the Spark SQL that provides all the RDD benefits with the optimized Spark SQL execution engine. It is defined as the distribution of collection of data. The Dataset API is available for Scala and Java. It is not available for Python, as the dynamic nature of Python provides the benefits of the Dataset API as a built–in feature.

DataFrame is a Dataset organized as named columns, which makes querying easy. Conceptually, the DataFrame is equivalent to a table in any relational database. The DataFrames can be created from a variety of sources like any structured data files, external relational data sources, or existing RDDs.

Spark Session

The entry point for all Spark SQL functionality is the Spark Session API. The Spark Session can be created using `SparkSession.builder()`.

```
import org.apache.spark.sql.SparkSession  // Line 1

val spark = SparkSession.builder()
                    .appName("PracticalSpark_SQL Application")
                    .getOrCreate()     // Line 2

import spark.implicits._                 // Line 3
```

In this code, line 3 is mandatory to enable all implicit conversions like converting RDDs to DataFrames.

Note Spark 2.0 provides the built-in support for Hive features to write queries using HiveQL and to read data from Hive tables.

In starting the Spark Shell, Spark Session will be created by default and it is not required to create the session manually again in the shell (see Figure 4-1).

```
Spark context Web UI available at http://10.0.2.15:4040
Spark context available as 'sc' (master = local[*], app id = local-1517568219555).
Spark session available as 'spark'.
Welcome to
      ____              __
     / __/__  ___ _____/ /__
    _\ \/ _ \/ _ `/ __/  '_/
   /___/ .__/\_,_/_/ /_/\_\   version 2.2.0
      /_/

Using Scala version 2.11.8 (Java HotSpot(TM) 64-Bit Server VM, Java 1.8.0_77)
Type in expressions to have them evaluated.
Type :help for more information.

scala>
```

Figure 4-1. *Spark Session in Spark Shell*

The details of Spark Shell were explained completely in Chapter 3.

Creating DataFrames

The DataFrames can be created by using existing RDDs, Hive tables, and other data sources like text files and external databases. The following example shows the steps to create DataFrames from the JSON file with SparkSession. The steps to create DataFrames from existing RDDs and other data sources is explained later in this chapter.
Add the contents in bookDetails.json as shown in Figure 4-2.

```
{"bookId":101, "bookName":"Practical Spark", "Author":"Dharanitharan G"}

{"bookId":102, "bookName":"Spark Core", "Author":"Subhashini R C"}

{"bookId":103, "bookName":"Spark SQL", "Author":"Dharanitharan G"}

{"bookId":104, "bookName":"Spark Streaming", "Author":" Subhashini R C "}
```

Figure 4-2. *bookDetails.json*

Follow the example shown here to create the DataFrame from the JSON content. Refer to Figure 4-3 for the output.

```
val bookDetails = spark.read.json("/home/SparkDataFiles/bookDetails.json")
```

```
scala> val bookDetails = spark.read.json("/home/SparkDataFiles/bookDetails.json")
bookDetails: org.apache.spark.sql.DataFrame = [Author: string, bookId: bigint ...

scala> bookDetails.show()
+---------------+------+---------------+
|         Author|bookId|       bookName|
+---------------+------+---------------+
|Dharanitharan G|   101|Practical Spark|
|   Subhashini RC|   102|     Spark Core|
|Dharanitharan G|   103|      Spark SQL|
|   Subhashini RC|   104|Spark Streaming|
+---------------+------+---------------+
```

Figure 4-3. *Creating DataFrame using JSON file*

The `spark.read.json("/filepath")` is used to read the content of the JSON file as a DataFrame. `bookDetails` is created as a DataFrame. The `show()` method is used to display the contents of a DataFrame in the `stdout`.

DataFrame Operations

DataFrame operations provides a structured data manipulation with APIs available in different languages such as Java, Scala, Python, and R. The DataFrames are the set of Dataset rows in Java and Scala.

The DataFrame operations are also called *Untyped transformations*. Shown here are examples of a few uptyped transformations available for DataFrames. It is recommended that you practice all the given examples. Refer to Figure 4-4 for the `printSchema()` function.

```
scala> bookDetails.printSchema()
root
 |-- Author: string (nullable = true)
 |-- bookId: long (nullable = true)
 |-- bookName: string (nullable = true)
```

Figure 4-4. *printSchema() function on a DataFrame*

The printSchema() function displays the schema of the DataFrame.

Untyped DataFrame Operation: Select

The select() transformation is used to select the required columns from the DataFrame. Refer to the following code and Figure 4-5.

```
bookDetails.select("bookId","bookName").show()
```

```
scala> bookDetails.select("bookId","bookName").show()
+------+---------------+
|bookId|       bookName|
+------+---------------+
|   101|Practical Spark|
|   102|     Spark Core|
|   103|      Spark SQL|
|   104|Spark Streaming|
+------+---------------+
```

Figure 4-5. *Untyped DataFrame operation: Select*

Untyped DataFrame Operation: Filter

The filter() transformation is used to apply the filter conditions on the DataFrame rows while retrieving the data. Refer to the following code for the filter operation and see Figure 4-6 for the output.

```
bookDetails.filter($"bookName" === "Spark Core").show()
```

```
scala> bookDetails.filter($"bookName" === "Spark Core").show()
+-------------+------+----------+
|       Author|bookId|  bookName|
+-------------+------+----------+
|Subhashini RC|   102|Spark Core|
+-------------+------+----------+
```

Figure 4-6. *Untyped DataFrame operation: Filter*

Note $"bookName" indicates the values of the column. Also, === (triple equal) must be used to match the condition.

Untyped DataFrame Operation: Aggregate Operations

The groupBy() transformation is used to apply filter aggregation on the DataFrame rows while retrieving the data. The following code shows the groupBy operation and Figure 4-7 displays the output.

```
val grouped = bookDetails.groupBy("Author")
val total = grouped.count()
```

```
scala> val grouped = bookDetails.groupBy("Author")
grouped: org.apache.spark.sql.RelationalGroupedDataset = org.apache.spa

scala> val total = grouped.count()
total: org.apache.spark.sql.DataFrame = [Author: string, count: bigint]

scala> total.show()
+---------------+-----+
|         Author|count|
+---------------+-----+
|Dharanitharan G|    2|
|  Subhashini RC|    2|
+---------------+-----+
```

Figure 4-7. *Untyped DataFrame operation: Aggregate operations*

These transformations can be chained together and written as

```
bookDetails.groupBy("Author").count().show()
```

Hint The equivalent SQL of these chained transformations is SELECT Author, COUNT(Author) FROM BookDetails GROUP BY Author.

Running SQL Queries Programatically

The `sql` function on the `SparkSession` allows us to run the SQL queries programmatically and it returns the DataFrame as a result.

Creating Views

It is necessary to create a view from the DataFrame to run the SQL queries directly based on the requirements. The views are always temporary and session scoped. They will be destroyed if the session that creates the views is terminated. There are two types of temporary views: temporary views and global temporary views. Assume that the view is like a relational database management system (RDBMS) view.

Once the view is created, the SQL query can be executed on the view by using the `sql` method in `SparkSession` as shown in Figure 4-8.

```
scala> bookDetails.createOrReplaceTempView("BookDetails")

scala> val rs =
     | spark.sql("SELECT Author,COUNT(Author) FROM BookDetails GROUP BY Author")
rs: org.apache.spark.sql.DataFrame = [Author: string, count(Author): bigint]

scala> rs.show()
+---------------+-------------+
|         Author|count(Author)|
+---------------+-------------+
|Dharanitharan G|            2|
|   Subhashini RC|            2|
```

Figure 4-8. *Running an SQL query programmatically, temporary view*

The function `createOrReplaceTempView()` is used to create a temporary view that is available only in the same `SparkSession`; that is, (`spark`).

The global temporary view can be created by using the `createGlobalTempView()` function. The global temporary view is shared among all the Spark sessions and remains alive until the Spark application is terminated. It is tied to the system preserved database `'global_temp'` and hence it is required to use a fully qualified table name like `global_temp.<table_name>` to refer it while using it in the query (see Figure 4-9).

```
scala> bookDetails.createGlobalTempView("BookDetails")

scala> val result = spark.sql("SELECT * FROM global_temp.BookDetails")
result: org.apache.spark.sql.DataFrame = [Author: string, bookId: bigint .

scala> result.show()
+---------------+------+---------------+
|         Author|bookId|       bookName|
+---------------+------+---------------+
|Dharanitharan G|   101|Practical Spark|
|  Subhashini RC|   102|     Spark Core|
|Dharanitharan G|   103|      Spark SQL|
|  Subhashini RC|   104|Spark Streaming|
+---------------+------+---------------+
```

Figure 4-9. *Running SQL query programmatically, global temporary view*

SPARK SQL EXERCISE 1: DATAFRAME OPERATIONS

1. Create the following data as `logdata.log` with comma delimiters as shown.

10:24:25,10.192.123.23,http://www.google.com/searchString,ODC1

10:24:21,10.123.103.23,http://www.amazon.com,ODC1

10:24:21,10.112.123.23,http://www.amazon.com/Electronics,ODC1

10:24:21,10.124.123.24,http://www.amazon.com/Electronics/storagedevices,ODC1

10:24:22,10.122.123.23,http://www.gmail.com,ODC2

10:24:23,10.122.143.21,http://www.flipkart.com,ODC2

10:24:21,10.124.123.23,http://www.flipkart.com/offers,ODC1

Note   The schema for these data is: `Time, IpAddress, URL, Location`

---

2. Create a DataFrame of the created log file using `spark.read.csv`.

Note   The `spark.read.csv` reads the data from a file with comma delimiters by default and the column names of the DataFrame would be _c0, _c1, and so on. The different data sources, options, and the format for creating DataFrames with different schema is discussed later in this chapter.

---

3. Create a temporary view named 'LogData'.

4. Create a global temporary view named `'LogData_Global'`. Observe the difference between the temporary view and global temporary view by executing the query with a temporary view in a different Spark session.

5. Write and run SQL queries programatically for the following requirements.

    •   How many people accessed the Flipkart domain in each location?

    •   Who accessed the Flipkart domain in each location? List their IpAddress.

    •   How many distinct Internet users are available in each location?

    •   List the unique locations available.

Dataset Operations

Datasets are like RDDs. Dataset APIs provide a type safe and object-oriented programming interface. The DataFrame is an alias for untyped `Dataset[Row]`. Datasets also provide high-level domain-specific language operations like `sum()`, `select()`, `avg()`, and `groupby()`, which makes the code easier to read and write.
Add the contents shown in Figure 4-10 to `BookDetails.json`.

```
{"bookId":101, "bookName":"Practical Spark", "Author":"Dharanitharan G"}

{"bookId":102, "bookName":"Spark Core", "Author":"Subhashini R C"}

{"bookId":103, "bookName":"Spark SQL", "Author":"Dharanitharan G"}

{"bookId":104, "bookName":"Spark Streaming", "Author":" Subhashini R C "}
```

Figure 4-10. `BookDetails.json`

Create a case class for the bookDetails schema as shown here. Figure 4-11 displays the result.

```
case class BookDetails (bookId:String, bookname: String, Author:String)
```

```
scala> case class BookDetails (bookId:String, bookname: String, Author:String)
defined class BookDetails
```

Figure 4-11. *Case class for BookDetails.json*

Now, create the DataSet by reading from the JSON file.

```
val bookDetails =     spark.read.json("/home/SparkDataFiles/bookDetails.
json").as[BookDetails]
```

This code creates the Dataset (named bookDetails) and it is represented as org.apache.spark.sql.Dataset[BookDetails] because the case class, BookDetails, is used to map the schema. See Figure 4-12 for the output.

```
scala> case class BookDetails (bookId:String, bookname: String, Author:String)
defined class BookDetails

scala> val inputPath = "/home/SparkDataFiles/bookDetails.json"
inputPath: String = /home/SparkDataFiles/bookDetails.json

scala> val bookDetails = spark.read.json(inputPath).as[BookDetails]
bookDetails: org.apache.spark.sql.Dataset[BookDetails] = [Author: string, book

scala> bookDetails.show()
+---------------+------+---------------+
|         Author|bookId|       bookName|
+---------------+------+---------------+
|Dharanitharan G|   101|Practical Spark|
|   Subhashini RC|   102|     Spark Core|
|Dharanitharan G|   103|      Spark SQL|
|   Subhashini RC|   104|Spark Streaming|
+---------------+------+---------------+
```

Figure 4-12. *Dataset operations*

It is possible to do all the DataFrame operations on Dataset as well.

Interoperating with RDDs

In Spark SQL, there are two methods for converting the existing RDDs into Datasets: the reflection-based approach and the programmatic interface.

Reflection-Based Approach to Infer Schema

The RDDs containing case classes can be automatically converted into a DataFrame using the Scala interface for Spark SQL. The case class defines the schema of the DataFrame. The column names of the DataFrames are read using the reflection from the names of the arguments of case classes. The RDD can be implicitly converted into a DataFrame and then converted into a table.

Add the following contents in bookDetails.txt:

```
101,Practical Spark,Dharanitharan G
102,Spark Core,Subhashini RC
103,Spark SQL,Dharanitharan G
104,Spark Streaming,Subhashini RC
```

Create a case class for the bookDetails schema.

```
case class BookDetails (bookId:String, bookname: String, Author:String)
```

Now, create an RDD from the bookDetails.txt file as shown in Figure 4-13.

```
scala> val inputPath = "/home/SparkDataFiles/bookDetails.txt"
inputPath: String = /home/SparkDataFiles/bookDetails.txt

scala> val bookDetails = sc.textFile(inputPath)
bookDetails: org.apache.spark.rdd.RDD[String] = /home/SparkDataFiles/bookDetails.txt
```

Figure 4-13. *Creating RDD from a file*

Note   sc is a SparkContext, which is available in a Spark Shell session.

Because the RDD is created from the text file, each element in the RDD is a string (each line in the file is converted as an element in the RDD).

Now the DataFrame can be created from the existing RDD bookDetails by using the toDF() function as shown in Figure 4-14. Observe that each element in the RDD is converted as a row in DataFrame and each field in the element is converted as a column.

```
scala> val bookDetailsDF = bookDetails.toDF()
bookDetailsDF: org.apache.spark.sql.DataFrame = [value: string]

scala> bookDetailsDF.show()
+--------------------+
|               value|
+--------------------+
|101,Practical Spa...|
|102,Spark Core,Su...|
|103,Spark SQL,Dha...|
|104,Spark Streami...|
+--------------------+


scala> bookDetailsDF.show(false)
+----------------------------------+
|value                             |
+----------------------------------+
|101,Practical Spark,Dharanitharan G|
|102,Spark Core,Subhashini RC       |
|103,Spark SQL,Dharanitharan G      |
|104,Spark Streaming,Subhashini RC  |
+----------------------------------+


scala> bookDetailsDF.printSchema()
root
 |-- value: string (nullable = true)
```

Figure 4-14. *Creating DataFrame from an existing RDD*

Because each element in the RDD contains only one field, there is only one column in the DataFrame. So, we need to create each element in the RDD with multiple fields as per requirements. Also, observe that the schema is inferred from the existing case class BookDetails. The column names of the DataFrame are taken from the names of arguments of the case class (see Figure 4-15).

```
scala> val details2 = bookDetails.map(line => line.split(","))
details2: org.apache.spark.rdd.RDD[Array[String]] = MapPartitionsRDD[11]

scala> val details_mapped = details2.map(x => BookDetails(x(0),x(1),x(2)))
details_mapped: org.apache.spark.rdd.RDD[BookDetails] = MapPartitionsRDD[12]

scala> val bookDetailsDF = details_mapped.toDF()
bookDetailsDF: org.apache.spark.sql.DataFrame

scala> bookDetailsDF.show(false)
+------+---------------+---------------+
|bookId|bookname       |Author         |
+------+---------------+---------------+
|101   |Practical Spark|Dharanitharan G|
|102   |Spark Core     |Subhashini RC  |
|103   |Spark SQL      |Dharanitharan G|
|104   |Spark Streaming|Subhashini RC  |
+------+---------------+---------------+
```

Figure 4-15. *Schema inference through reflection from case class attributes*

Now, the DataFrame can be registered as the temporary table and SQL queries can be run programmatically.

The schema can be represented by a StructType matching the structure of rows in the RDD created from the text file. Then, apply the schema to the RDD of rows via the createDataFrame method provided by the Spark Session.

```
import org.apache.spark.sql.types._
import org.apache.spark.sql._
```

These two imports are mandatory because the StructField and StructType should be used for creating the schema (see Figure 4-16).

```
scala> import org.apache.spark.sql._
import org.apache.spark.sql._

scala> import org.apache.spark.sql.types._
import org.apache.spark.sql.types._

scala> val schema = Array("bookID","bookName","AuthorName")
schema: Array[String] = Array(bookID, bookName, AuthorName)

scala> val fields = schema.map( x => StructField(x, StringType, nullable = true))
fields: Array[org.apache.spark.sql.types.StructField]

scala> val dfSchema = StructType(fields)
dfSchema: org.apache.spark.sql.types.StructType
```

Figure 4-16. *Schema creation using StructType*

Now, the created schema can be merged with the RDD as shown in Figure 4-17.

```
scala> val inputPath = "/home/SparkDataFiles/bookDetails.txt"
inputPath: String = /home/SparkDataFiles/bookDetails.txt

scala> val bookDetails = sc.textFile(inputPath)
bookDetails: org.apache.spark.rdd.RDD[String]

scala> val bookDetails2 = bookDetails.map(x => x.split(","))
bookDetails2: org.apache.spark.rdd.RDD[Array[String]]

scala> val fieldsMap = bookDetails2.map(x=>Row(x(0),x(1),x(2)))
fieldsMap: org.apache.spark.rdd.RDD[org.apache.spark.sql.Row]

scala> val bookDetailsDF = spark.createDataFrame(fieldsMap,dfSchema)
bookDetailsDF: org.apache.spark.sql.DataFrame

scala> bookDetailsDF.show(false)
+------+---------------+---------------+
|bookID|bookName       |AuthorName     |
+------+---------------+---------------+
|101   |Practical Spark|Dharanitharan G|
|102   |Spark Core     |Subhashini RC  |
|103   |Spark SQL      |Dharanitharan G|
|104   |Spark Streaming|Subhashini RC  |
+------+---------------+---------------+
```

Figure 4-17. *Programmatically specifying schema*

Different Data Sources

Spark SQL supports a variety of data sources like json, csv, txt, parquet, jdbc, and orc. In this module, we discuss the generic load and save functions and manually specifying options for loading and saving. It is also possible to run the SQL queries programatically directly on the files without creating the RDDs and DataFrames.

Generic Load and Save Functions

The default data source is parquet files, but the default can be configured by changing the spark.sql.sources.default property. See Figure 4-18 for how to use generic load and save functions.

```scala
scala> val inputPath = "/home/SparkDataFiles/users.parquet"
inputPath: String = /home/SparkDataFiles/users.parquet

scala> val userDetailsDF = spark.read.load(inputPath)
userDetailsDF: org.apache.spark.sql.DataFrame

scala> val userName = userDetailsDF.select("name")
userName: org.apache.spark.sql.DataFrame = [name: string]

scala> userName.show(false)
+------+
|name  |
+------+
|Alyssa|
|Ben   |
+------+

scala> userName.write.save("/home/SparkDataFiles/UserNames.parquet")
```

Figure 4-18. *Generic load and save functions*

If the property spark.sql.sources.default is not changed, the type of data source can be specified manually as explained later.

Manually Specifying Options

The format of the data sources can be manually specified by using the `format()` function, as shown in Figure 4-19.

```scala
scala> val inputPath = "/home/SparkDataFiles/bookDetails.json"
inputPath: String = /home/SparkDataFiles/bookDetails.json

scala> val bookDetailsDF = spark.read.format("json").load(inputPath)
bookDetailsDF: org.apache.spark.sql.DataFrame

scala> val bookNames = bookDetailsDF.select("bookName","Author")
bookNames: org.apache.spark.sql.DataFrame

scala> bookNames.show(false)
+---------------+---------------+
|bookName       |Author         |
+---------------+---------------+
|Practical Spark|Dharanitharan G|
|Spark Core     |Subhashini RC  |
|Spark SQL      |Dharanitharan G|
|Spark Streaming|Subhashini RC  |
+---------------+---------------+

scala> bookNames.write.format("csv").save("/home/SparkDataFiles/BookNames")
```

Figure 4-19. *Manually specifying options for loading and saving files*

To create parquet files, the format can be specified as `parquet` for the `save` function.

Run SQL on Files Directly

The SQL queries can be run directly on the files programmatically instead of using load functions, as shown in Figure 4-20. Use the created `bookDetails.parquet` file as the input file.

```
scala> val sqlDF = spark.sql("SELECT * FROM parquet.`/home/SparkDataFiles/bookDetails.parquet`")
sqlDF: org.apache.spark.sql.DataFrame = [bookName: string, Author: string]

scala> sqlDF.show()
+----------------+----------------+
|        bookName|          Author|
+----------------+----------------+
|Practical Spark|Dharanitharan G|
|      Spark Core|  Subhashini RC|
|       Spark SQL|Dharanitharan G|
|Spark Streaming|  Subhashini RC|
+----------------+----------------+
```

Figure 4-20. *Running SQL on files directly (parquet source)*

The same can be done for a json data source, shown in Figure 4-21.

```
scala> val sqlDF = spark.sql("SELECT * FROM json.`/home/SparkDataFiles/bookDetails.json`")
sqlDF: org.apache.spark.sql.DataFrame = [Author: string, bookId: bigint ... 1 more field]

scala> sqlDF.show()
+----------------+------+----------------+
|          Author|bookId|        bookName|
+----------------+------+----------------+
|Dharanitharan G|   101|Practical Spark|
|  Subhashini RC|   102|      Spark Core|
|Dharanitharan G|   103|       Spark SQL|
|  Subhashini RC|   104|Spark Streaming|
+----------------+------+----------------+
```

Figure 4-21. *Running SQL on files directly (json source)*

Spark SQL automatically infers the schema of json files and loads them as
Dataset[Row].

131

JDBC to External Databases

Spark SQL allows users to connect to external databases through JDBC (i.e., Java DataBase Connectivity) connectivity. The tables from the databases can be loaded as DataFrame or Spark SQL temporary tables using the Datasources API.

The following properties are mandatory to connect to the database.

- *URL:* The JDBC URL to connect to (e.g., jdbc:mysql://${jdbcHostname}:${jdbcPort}/${jdbcDatabase}).

- *Driver:* The class name of the JDBC driver to connect to the URL (e.g., com.mysql.jdbc.Driver, for mysql database).

- *UserName and Password:* To connect to the database.

The following code creates the DataFrame from the mysql table.

```
val jdbcDF = spark.read.format("jdbc")
                    .option("url", "jdbc:mysql:localhost:3306/sampleDB")
                    .option("dbtable", "sampleDB.bookDetailsTable ")
                    .option("user", "<username>")
                    .option("password", "<password>")
                    .load()
```

It is mandatory to keep the jar for mysql database in the Spark classpath.

The same can be done by specifying the connection properties separately and using the same with direct read as shown here where, spark is the Spark Session.

```
val connectionProperties = new Properties()
connectionProperties.put("user", "username")
connectionProperties.put("password", "password")

val jdbcDF2 = spark.read.jdbc("jdbc:mysql:localhost:3306/sampleDB",
                            "schema.tablename", connectionProperties)
```

Spark SQL allows the users to write into the external tables of any databases. This code can be used to write the data in a mysql table.

```
jdbcDF.write.format("jdbc")
           .option("url", " jdbc:mysql:localhost:3306/sampleDB")
           .option("dbtable", "schema.tablename")
           .option("user", "username")
           .option("password", "password")
           .save()
```

Working with Hive Tables

Spark SQL supports reading and writing data stored in Apache Hive. Configuration of Hive is done by placing hive-site.xml in the configuration folder of Spark. When it is not configured by hive-site.xml, Spark automatically creates metastore db in the current directory, which defaults to the spark-warehouse directory in the current directory when the Spark application is started.

To work with Hive, instantiate SparkSession with Hive support as shown in the following code. Refer to Figure 4-22.

```
import org.apache.spark.sql.Row
import org.apache.spark.sql.SparkSession

val spark = SparkSession.builder().appName("Spark Hive Example").
config("spark.sql.warehouse.dir", "/home/Spark").enableHiveSupport().
getOrCreate()
```

```
scala> import org.apache.spark.sql.Row
import org.apache.spark.sql.Row

scala> import org.apache.spark.sql.SparkSession
import org.apache.spark.sql.SparkSession

scala>

scala> val spark = SparkSession.builder().appName("Spark Hive Example").config("spark.
sql.warehouse.dir", "/home/Spark").enableHiveSupport().getOrCreate()
18/08/17 04:57:18 WARN SparkSession$Builder: Using an existing SparkSession; some conf
iguration may not take effect.
spark: org.apache.spark.sql.SparkSession = org.apache.spark.sql.SparkSession@2b4954a4

scala>
```

Figure 4-22. *SparkSession with Hive support*

Now, you can create a Hive table as shown in the following code. The output is shown in Figure 4-23.

```
scala> case class authors(name:String, publisher: String)
defined class authors

scala> sql("CREATE TABLE IF NOT EXISTS authors (name String, publisher String) ROW FOR
MAT DELIMITED FIELDS TERMINATED BY ',' ")
18/08/17 05:06:44 WARN HiveMetaStore: Location: file:/home/vagrant/spark-warehouse/aut
hors specified for non-external table:authors
res0: org.apache.spark.sql.DataFrame = []

scala> sql("LOAD DATA LOCAL INPATH '/home/Spark/authors.txt' INTO TABLE authors")
res1: org.apache.spark.sql.DataFrame = []

scala> sql("SELECT * FROM authors").show()
+-------------+---------+
|         name|publisher|
+-------------+---------+
|    Subhashini|   Apress|
|Dharanitharan|   Apress|
+-------------+---------+
```

Figure 4-23. *Working with Hive table*

case class authors(name:String, publisher: String)

```
sql("CREATE TABLE IF NOT EXISTS authors (name String, publisher String) ROW
FORMAT DELIMITED FIELDS TERMINATED BY ',' ")
sql("LOAD DATA LOCAL INPATH '/home/Spark/authors.txt' INTO TABLE authors")

sql("SELECT * FROM authors").show()
```

The table authors can be viewed in Hive as shown in Figure 4-24.

```
hive> show tables;
OK
authors
Time taken: 0.038 seconds, Fetched: 1 row(s)
hive> select * from authors;
OK
Subhashini          Apress
Dharanitharan       Apress
Time taken: 1.833 seconds, Fetched: 2 row(s)
```

Figure 4-24. *authors table in hive prompt*

Building Spark SQL Application with SBT

The SBT installation procedure was already discussed in the previous chapter. Follow the further steps here to add the SparkSQL dependencies in the build.sbt file. Add the content shown here to the build.sbt file.

```
name := "SparkSQL-DemoApp"
version := "1.0"
scalaVersion := "2.11.8"
libraryDependencies += "org.apache.spark" % "spark-core_2.11" % "2.1.0"
libraryDependencies += "org.apache.spark" % "spark-sql_2.11" % "2.1.0"
```

SBT downloads the required dependencies for the Spark SQL and keeps it in the local repository if it is not available while building the jar.

Note   Any other build tools like maven can also be used to build the package, the SBT is recommended for packaging the Scala classes.

Let's write a Spark SQL application to display or get the list of books written by author "Dharanitharan G" from the bookDetails.json file.

Create a Scala file named BooksByDharani.scala and add the following code:

```scala
import org.apache.spark.sql._
import org.apache.spark.sql.SparkSession

object BooksByDharani {
 def main(args:Array[String]) :Unit = {
    val spark = SparkSession.builder()
                          .appName("BooksByDharani")
                          .getOrCreate()
    import spark.implicits._
    val bookDetails = spark.read.json(args(0))
    bookDetails.createGlobalTempView("BookDetails")
    val result = spark.sql("SELECT * FROM global_temp.BookDetails")
    result.rdd.saveAsTextFile(args(1))
 }
}
```

The input path and the output path are specified as args(0) and args(1) to pass it as command-line arguments while submitting it to the cluster.

It is mandatory to import spark.implicits._ as discussed at the beginning of this chapter to enable the implicit conversions of DataFrames from RDD.

Create the folder structure shown in Figure 4-25, where BooksByDharani is the folder and src/main/scala are subfolders.

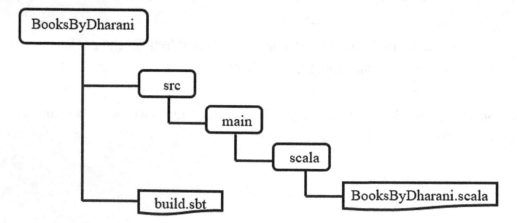

Figure 4-25. *Folder structure*

Navigate to the folder BooksByDharani (i.e., cd /home/BooksByDharani). Now execute the Scala build package command to build the jar file.

```
> cd /home/BooksByDharani
> sbt clean package
```

Once the build has succeeded, it creates the project and target directory shown in Figure 4-26.

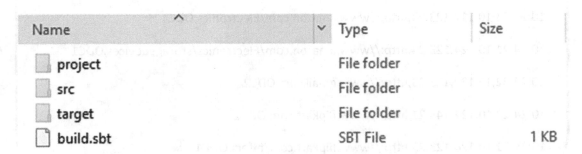

Name	Type	Size
project	File folder	
src	File folder	
target	File folder	
build.sbt	SBT File	1 KB

Figure 4-26. *SBT build directory structure*

SBT creates the application jar SparkSQL-DemoApp-1.0_2.11.jar in the target directory. Now, the application can be submitted to the Spark cluster by using the following command.

```
spark-submit --class BooksByDharani --master spark://<hostIP>:<port>
SparkSQL-DemoApp-1.0_2.11.jar <inputfilepath> <outputfilepath>
```

where spark://<hostIP>:<port> is the URI for the Spark master. By default, the Spark master runs on port 7077. However, that can be changed in the configuration files.

SPARK SQL EXERCISE 2: DATAFRAME OPERATIONS

1. Create the following data as `logdata.log` with comma delimiters as shown.

```
10:24:25,10.192.123.23,http://www.google.com/searchString,ODC1

10:24:21,10.123.103.23,http://www.amazon.com,ODC1

10:24:21,10.112.123.23,http://www.amazon.com/Electronics,ODC1

10:24:21,10.124.123.24,http://www.amazon.com/Electronics/storagedevices,ODC1

10:24:22,10.122.123.23,http://www.gmail.com,ODC2

10:24:23,10.122.143.21,http://www.flipkart.com,ODC2

10:24:21,10.124.123.23,http://www.flipkart.com/offers,ODC1
```

Note   The schema for these data is `Time, IpAddress, URL, Location`.

---

2. Create an RDD from the created file with the column names as specified by using the schema inference through reflection method.

3. Create a DataFrame from the created RDD and register it as a global temporary view named `LogDetails_Global`.

4. Write a SQL query to find the number of unique IP addresses in each location.

5. Save the DataFrame created in Question 3 as a `json` file, using the Spark `write` method by specifying the `json` format.

6. Run the same SQL query to find the number of unique IP addresses in each location directly on the `json` file created without creating a DataFrame.

Points to Remember

- Spark SQL is the Spark module for processing structured data.

- DataFrame is a Dataset organized as named columns, which makes querying easy. It is conceptually equivalent to a table in a relational database or a data frame in R/Python, but with richer optimizations under the hood.

- Dataset is a new interface added in Spark SQL that provides all the RDD benefits with the optimized Spark SQL execution engine.

In the next chapter, we are going to discuss how to work with Spark Streaming.

Introduction to Spark Streaming

In Chapter 4 we discussed how to process structured data using DataFrames, Spark SQL, and Datasets.

The recommended background for this chapter is some prior experience with Scala. In this chapter, we are going to focus on real-time processing using Apache Spark. We will be focusing on these areas:

- Data processing.

- Streaming data.

- Why streaming data are important.

- Introduction to Spark Streaming.

- Spark Streaming example using TCP Socket.

- Stateful streaming.

- Streaming application considerations.

141

S. Chellappan and D. Ganesan, *Practical Apache Spark*, https://doi.org/10.1007/978-1-4842-3652-9_5

Data Processing

Data can be processed in two ways.

- *Batch processing:* A group of transactions are collected over a period of time and are processed as a one single unit of work or by dividing it into smaller batches. Batch processing gives insight about what happened in past. Examples include payroll and billing systems.

- *Real-time processing:* Data are processed as and when they are genearted. Real-time processing gives insight about what is happening now. An example is bank ATMs.

Streaming Data

Data that are generated continuously by different sources are known as streaming data. These data need to be processed incrementally to get insight about what is happening now. The stream data could be any of the following:

- Web clicks.

- Website monitoring.

- Network monitoring.

- Advertising.

Why Streaming Data Are Important

Streaming data are important because:

- Tracking of web clicks can be used to recommend a relevant product to a user.

- Tracking of logs could help to understand the root cause of the failure.

Introduction to Spark Streaming

Spark Streaming is an extension of the core Spark API. Spark Streaming captures continuous streaming data and process data in near real time. Near real time means that Spark does not process data in real time, but instead processes data in microbatches, in just a few milliseconds.

There are some of the notable features of Spark Streaming:

- Scalable, high-throughtput, and fault-tolerant stream processing.

- Data can be ingested from different sources such as TCP sockets, Kafka, and HDFS/S3.

- Data can be processed using high-level Spark Core APIs such as map, join, and window.

- Scala, Java, and Python APIs support.

- Final results can be stored in HDFS, databases, and dashboards.

Figure 5-1 illustrates the Spark Streaming architecture.

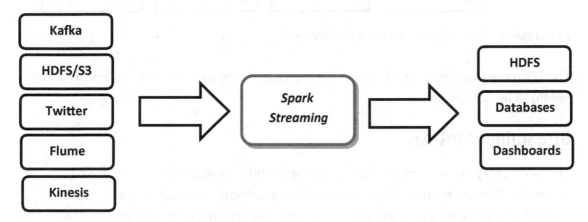

Figure 5-1. *Spark Streaming architecture*

Internal Working of Spark Streaming

Spark Streaming captures live input data streams and divides the streaming data into batches. These batches are processed by the Spark Streaming engine to generate the final stream results. This Spark Streaming process is illustrated in Figure 5-2.

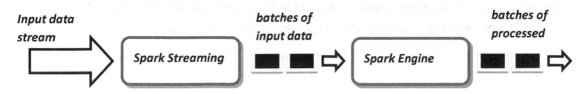

Figure 5-2. *Internal workings of Spark Streaming*

Spark Streaming Concepts

Let's discuss the some of the basic concepts of Spark Streaming.

Discretized Streams (DStream)

The basic abstraction provided by Spark Streaming is DStream. The DStream is a representation of a continuous series of RDDs as shown in Figure 5-3. Each RDD in a DStream contains data from a certain interval.

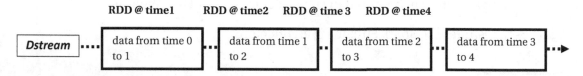

Figure 5-3. *Discretized streams (DStream)*

The DStream can be obtained from different sources. For example, a text stream can be obtained from TCP socket.

Streaming Context

The main entry point for Spark Streaming applications is Streaming Context. The Streaming Context is equivalent to SparkContext in Spark. Streaming Context can be configured in the same way as Spark Context, but it requires batch durations in milliseconds, seconds, or minutes.

DStream Operations

The RDD operations can be applied to each batch to process a continuous stream of data (e.g., map, flatMap, filter, etc.). There are two types of RDD operations: transformations and output operations. Transformations create a new DStream from an existing DStream. Output operations write data to a file system.

Two important streaming methods are `start`, which starts the execution of DStreams, and `awaitTermination`, which waits for computation to terminate.

Spark Streaming Example Using TCP Socket

Let's discuss how we can perform the task of counting the occurrences of a word in text data received from a data server listening on a TCP Socket. Refer to the following code.

```
package com.apress.book

import org.apache.spark.sql.{Row, SparkSession}
import org.apache.spark.streaming.{Seconds, StreamingContext}
import org.apache.spark.storage.StorageLevel

object SparkWordCountStreaming{

  def main(args: Array[String])
  {

    // Create Spark Session and Spark Context

    val spark = SparkSession.builder.appName(getClass.getSimpleName).
    getOrCreate()

    // Get the Spark context from the Spark session to create streaming
    context

    val sc = spark.sparkContext

    // Create the streaming context, interval is 40 seconds

    val ssc = new StreamingContext(sc, Seconds(40))

    // Set the check point directory to save the data to recover when
    there is a crash

    ssc.checkpoint("/tmp")
// Create a DStream that connects to hostname:port to stream data from a
   TCP source.

      // Set the StorageLevel as StorageLevel.MEMORY_AND_DISK_SER which
         indicates that the data will be stored in memory and if it
         overflows, in disk as well
```

```
    val lines = ssc.socketTextStream("localhost", 9999, StorageLevel.
    MEMORY_AND_DISK_SER)
```

```
// count the number of words in text data received from a data server
    listening on a TCP socket.
```

```
// Split each line into words
```

```
    val words = lines.flatMap(_.split(" "))
```

```
// Count each word in each batch
```

```
    val pairs = words.map(word => (word, 1))
    val wordCounts = pairs.reduceByKey(_ + _)
```

```
// Print the elements of each RDD generated in this DStream to the
    console
```

```
    wordCounts.print()
```

```
// Start streaming
```

```
    ssc.start()
```

```
// Wait until the application is terminated
```

```
    ssc.awaitTermination()
  }
}
```

Build SparkWordCountStreaming.scala using SBT. The folder structure is shown in Figure 5-4.

Name	Date modified	Type	Size
src	30-03-2018 11:50	File folder	
build.sbt	30-03-2018 15:26	SBT File	1 KB

Figure 5-4. *SparkWordCountStreaming folder structure*

The build.sbt file is shown here.

```
name := "Spark_Streaming_WordCount"

version := "1.0"

scalaVersion := "2.11.8"

val sparkVersion = "2.1.0"

libraryDependencies ++= Seq(
  "org.apache.spark" %% "spark-core" % sparkVersion,
  "org.apache.spark" %% "spark-sql" % sparkVersion,
  "org.apache.spark" %% "spark-streaming" % sparkVersion
  )
```

Next, navigate to the corresponding folder and type sbt clean package as shown in Figure 5-5 to build the project.

```
c:\SparkStreamingDemos\Spark_Streaming_Wordcount_Demo>sbt clean package
```

Figure 5-5. *Command to build SparkStreamingWordCount project*

Next, run Netcat, a small utility found in most Unix-like systems, as a data server (see Figure 5-6).

```
        nc -lk 9999

Welcome to Apress Pulication
```

Figure 5-6. *Running Netcat utility*

Next, copy the executable `spark_streaming_wordcount_2.11-1.0.jar` to the node where the Spark cluster is running and use the Spark `submit` command to submit the `SparkStreamingWordCount` application to the Spark cluster as shown here. The output is shown in Figure 5-7.

```
-------------------------------------------
Time: 1522595920000 ms
-------------------------------------------
(Welcome,1)
(Publication,1)
(Apress,1)
(to,1)
```

Figure 5-7. *Streaming output starts 40 seconds after* `ssc.start`

spark-submit --class com.apress.book. SparkWordCountStreaming --master spark://
localhost:7077 /home/data/spark_streaming_wordcount_2.11-1.0.jar
Figure 5-8 displays the streaming output 40 seconds later.

```
-------------------------------------------
Time: 1522596240000 ms
-------------------------------------------
(Publiations,1)
(Publication,1)
(Springer,1)
(Apress,1)
(Book,1)
```

Figure 5-8. *Streaming output 40 seconds later*

The streaming process continues until termination of the source.

Note   Here, word count computation is performed on each RDD based on the specified interval.

Stateful Streaming

The Spark Streaming architecture is a microbatch architecture. The incoming data are grouped into microbatches called DStream. DStream represents a continuous series of RDDs. When data are tracked on each RDD, this is known as stateless streaming. The previous example is an example for stateless streaming. When data are tracked across, it is known as stateful streaming.

There are two types of stateful streaming.

- Window-based tracking.

- Full-session-based tracking.

Window-Based Streaming

Apache Spark provides windowed computations that can be used to perform transformations over a sliding window of data. Window operation requires two parameters.

- *Window length*: The duration of the window.

- *Window interval*: The interval at which window operation is performed.

For example, if the window interval is 3 seconds and slide interval is 2 seconds, computations will be performed every 2 seconds on the batches that have arrived in the last 3 seconds.

In Figure 5-9, the incoming batches are grouped every 3 units of time (window interval) and the computations are done every 2 units of time (slide interval).

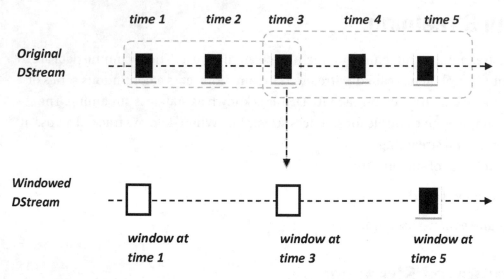

Figure 5-9. *Window operations*

Let's explore window operations with an example. We extend the earlier word count example by computing word counts over the last 30 seconds of data, every 10 seconds. This is achieved using the reduceByKeyAndWindow transformation. Refer to the following code.

```
package com.apress.book

import org.apache.spark.sql.{Row, SparkSession}
import org.apache.spark.streaming.{Seconds, StreamingContext}
import org.apache.spark.storage.StorageLevel

object WordCountByWindow{
  def main(args: Array[String])
  {

    // Create the Spark Session and the Spark Context
          val spark = SparkSession
                .builder
                .appName(getClass.getSimpleName)
                .getOrCreate()

    // Get the Spark Context from the Spark Session to create streaming
      context
          val sc = spark.sparkContext
```

```
// Create the streaming context, interval is 10 seconds
    val ssc = new StreamingContext(sc, Seconds(10))
```

```
// Set the checkpoint directory to save the data to recover when
   there is a crash
    ssc.checkpoint("/tmp")
```

```
// Create a DStream that connects to hostname:port to stream data
   from a TCP source.
```

```
// Set the StorageLevel as StorageLevel.MEMORY_AND_DISK_SER which
   indicates that the data will be stored in memory and if it
   overflows, in disk as well
```

```
// count the number of words in text data received from a data
   server listening on a TCP socket.
```

```
 val lines = ssc.socketTextStream("localhost",9999, StorageLevel.
 MEMORY_AND_DISK_SER)
```

```
// Split each line into words
 val words = lines.flatMap(_.split(" "))
// Count each word in over the last 30 seconds of data
 val pairs = words.map(word => (word, 1))
    val wordCounts = pairs.reduceByKeyAndWindow((x: Int, y: Int) =>
    x+y, Seconds(30), Seconds(10))
wordCounts.print()
```

```
// Start the streaming
    ssc.start()
// Wait until the application is terminated
    ssc.awaitTermination()
  }
}
```

Build the project and submit the WordCountByWindow application to the Spark Cluster. The Streaming data are shown in Figures 5-10 and 5-11, and the output is shown in Figure 5-12.

```
Welcome to Apress Publications
```

Figure 5-10. *Text data*

```
Springer Conference
```

Figure 5-11. *Text data*

```
-------------------------------------------
Time: 1522609540000 ms
-------------------------------------------
(Welcome,1)
(Apress,1)
(to,1)
(Publications,1)
```

Figure 5-12. *Word count over last 30 seconds*

Figure 5-13 displays the word count over the last 30 seconds.

```
-------------------------------------------
Time: 1522609550000 ms
-------------------------------------------
(Welcome,1)
(Springer,1)
(Apress,1)
(Conference,1)
(to,1)
(Publications,1)
```

Figure 5-13. *Word count over last 30 seconds*

Full-Session-Based Streaming

When data are tracked starting from the streaming job, this is known as full-session-based tracking. In full-session-based tracking, checking the previous state of the RDD is necessary to update the new state of the RDD.

Let's explore full-session-based tracking with an example. We extend the earlier word count program to count each word starting from the streaming job. This is achieved with the help of updateStateByKey (see Figure 5-14).

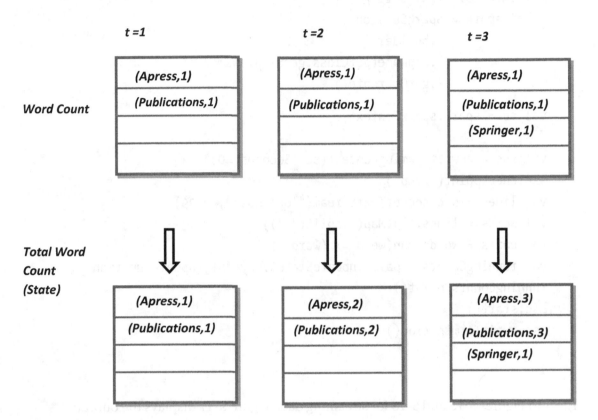

Figure 5-14. *Word count starting from the streaming job*

The code for full-session-based word count program is given here.

```
package com.apress.book

import org.apache.spark.sql.{Row, SparkSession}
import org.apache.spark.streaming.{Seconds, StreamingContext}
import org.apache.spark.storage.StorageLevel

object UpdateStateByKeyWordCount{

  def updateFunction(newValues: Seq[Int], runningCount: Option[Int]):
  Option[Int] = {
   val newCount = runningCount.getOrElse(0) + newValues.sum
```

```scala
      Some(newCount)
    }

  def main(args: Array[String]) {
      val spark = SparkSession
                    .builder
                    .appName(getClass.getSimpleName)
                    .getOrCreate()

    val sc = spark.sparkContext

    val ssc = new StreamingContext(sc, Seconds(40))
    ssc.checkpoint("/tmp")
    val lines = ssc.socketTextStream("localhost", 9999)
    val words = lines.flatMap(_.split(" "))
    val pairs = words.map(word => (word, 1))
    val runningCounts = pairs.updateStateByKey[Int](updateFunction _)
    runningCounts.print()
    ssc.start()
    ssc.awaitTermination()
  }
}
```

Refer to Figures 5-15 and 5-16 for streaming data. Figure 5-17 displays the output.

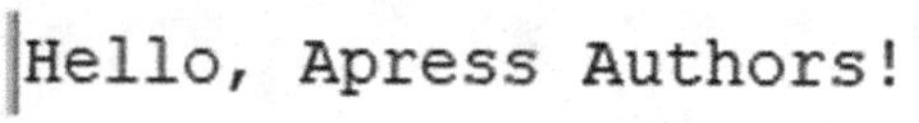

```
Hello, Apress Authors!
```

Figure 5-15. *Text data*

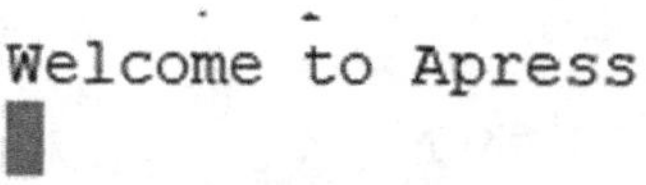

```
Welcome to Apress
```

Figure 5-16. *Text data*

```
----------------------------------------------
Time: 1522614640000 ms
----------------------------------------------
(Hello,,1)
(Apress,1)
(Authors!,1)
```

Figure 5-17. *Word count after 40 seconds*

Figure 5-18 shows the word count after 80 seconds.

```
----------------------------------------------
Time: 1522614720000 ms
----------------------------------------------
(Welcome,1)
(Hello,,1)
(Apress,2)
(Authors!,1)
(to,1)
```

Figure 5-18. *Word count after 80 seconds*

Streaming Applications Considerations

Spark Streaming applications are long-running applications that accumulate metadata over time. It is therefore necessary to use checkpoints when you perform stateful streaming. The checkpoint directory can be enabled using the following syntax.

```
ssc.checkpoint(directory)
```

Note   In Spark Streaming, the Job tasks are load balanced across the worker nodes automatically.

Points to Remember

- Spark Streaming is an extension of Spark Core APIs.

- The Spark Streaming architecture is a microbatch architecture.

- DStreams represents a continuous stream of data.

- The entry point for the streaming application is Streaming Context.

- RDD operations can be applied to microbatches to process data.

In the next chapter, we will be discussing how to work with Spark Structure Streaming.

CHAPTER 6

Spark Structured Streaming

In the previous chapter, you learned the concepts of Spark Streaming and stateful streaming. In this chapter, we are going to discuss structured stream processing built on top of the Spark SQL engine.

The recommended background for this chapter is to have some prior experience with Scala. Some familiarity with the command line is beneficial. The mandatory prerequisite for this chapter is completion of the previous chapters assuming that you have practiced all the demos.

In this chapter, we are going to discuss structured stream processing built on top of the Spark SQL engine. In this chapter, we are going to focus on the following topics:

- What Spark Structured Streaming is.

- Spark Structured Streaming programming model.

- Word count example using Structured Streaming.

- Creating streaming DataFrames and streaming Datasets.

- Operations on streaming DataFrames and Datasets.

- Stateful Structured Streaming, including window operation and watermarking.

- Triggers.

- Fault tolerance.

157

© Subhashini Chellappan, Dharanitharan Ganesan 2018
S. Chellappan and D. Ganesan, *Practical Apache Spark*, https://doi.org/10.1007/978-1-4842-3652-9_6

What Is Spark Structured Streaming?

Spark Structured Streaming is a fault-tolerant, scalable stream processing engine built on top of Spark SQL. The computations are executed on an optimized Spark SQL engine. The Scala, Java, R, or Python Dataset/DataFrame API is used to express streaming computation. Structured Streaming provides fast, scalable, fault-tolerant, end-to-end, exactly-once stream processing. Spark internally processes Structured Streaming queries using a microbatch processing engine. The process streams data as a series of small batch jobs.

Spark Structured Streaming Programming Model

The new stream processing model treats live data streams as a table that is being continuously appended. The streaming computations are expressed as a batch-like query and Spark runs this as an incremental query on the unbounded input table. In this model, the input data stream is considered as the *input table*. Every data item that is coming from the stream is considered a new row being appended to the input table (see Figure 6-1).

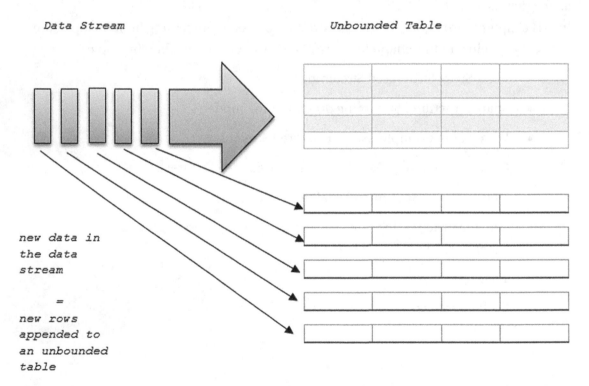

Figure 6-1. *Structured Streaming programming model*

A query on the input table will generate a *result table*. At every trigger interval, a new row is appended to the input table and this eventually updates the result table. We need to write the result rows to the external sink whenever the result table is updated (see Figure 6-2).

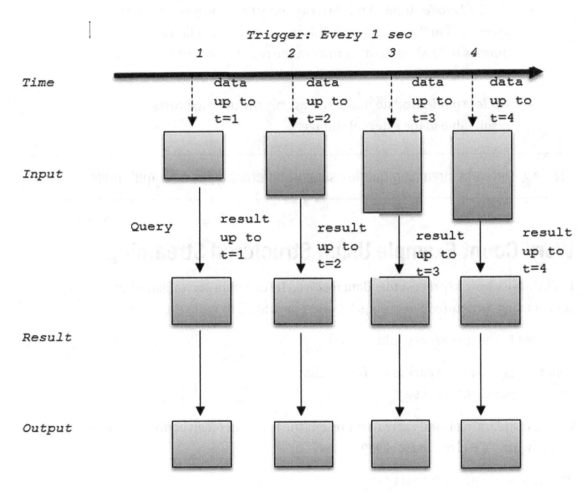

Figure 6-2. Programming model for Structured Streaming

Here, the output denotes what we need to write to the external storage. There are three different modes to specify output.

- *Complete mode*: Complete mode writes the entire result table to the external storage.

- *Append mode*: Append mode writes only the new rows that are appended to the result table. This mode can be applied on the queries only when existing rows in the result table are not expected to change.

- *Update mode*: Update mode writes only the updated rows in the result table to the external storage.

Note   Different streaming queries support different types of output mode.

Word Count Example Using Structured Streaming

Let's discuss how to process text data received from a data server listening on a TCP socket using Structured Streaming. We use Spark Shell to write the code.

```
// import the necessary classes.

import org.apache.spark.sql.functions._
import spark.implicits._
```

Create a streaming DataFrame to represent the text data received from a data server listening on TCP (localhost:9999).

```
val lines = spark.readStream
  .format("socket")
  .option("host", "localhost")
  .option("port", 9999)
  .load()
```

Transform the DataFrame to count the occurrences of a word that is received from the data server.

```
// Convert line DataFrame into Dataset and split the lines into multiple words
val words = lines.as[String].flatMap(_.split(" "))
// Generate running word count
val wordCounts = words.groupBy("value").count()
```

> **Note** Because we are using Spark Shell to run the code, there is no need to create Spark Session. Spark Session and Spark Context will be available by default.

Here, DataFrame represents an unbounded table. This unbounded table contains one column of string named value. Each line in the streaming data represents a row in the unbounded table.

```
// Write a query to print running counts of the word to the console
val query = wordCounts.writeStream
  .outputMode("complete")
  .format("console")
  .start()
query.awaitTermination()
```

Start the Netcat server using the following command and type a few messages as shown in Figures 6-3 and 6-4.

```
netcat -lk 9999
```

```
Hello Authors!
```

Figure 6-3. *Text data*

```
Welcome to Apress Publications!
```

Figure 6-4. *Text data*

The running counts of the words are shown in Figures 6-5 and 6-6. The query object handles the active streaming data until the application is terminated.

```
scala> ------------------------------------------
Batch: 0
------------------------------------------
+--------+-----+
|   value|count|
+--------+-----+
|Authors!|    1|
|   Hello|    1|
+--------+-----+
```

Figure 6-5. *Running word count*

```
scala> ------------------------------------------
Batch: 1
------------------------------------------
+-------------+-----+
|        value|count|
+-------------+-----+
|     Authors!|    1|
|        Hello|    1|
|       Apress|    1|
|Publications!|    1|
|      Welcome|    1|
|           to|    1|
+-------------+-----+
```

Figure 6-6. *Running word count*

Structured Streaming keeps only the minimal intermediate state data that are required to update the state. In this example, Spark keeps the intermediate count of word.

Creating Streaming DataFrames and Streaming Datasets

The `SparkSession.readStream()` returns the `DataStreamReader` interface. This interface is used to create streaming DataFrames. We can also specify the source: data format, schema, options, and so on.

The following are the built-in input sources.

- *File source*: Reads files written in a directory as a stream of data. The supported file formats are `text`, `csv`, `json`, `orc`, and `parquet`.

- *Kafka source*: Reads data from Kafka.

- *Socket source*: Reads UTF8 text data from a socket connection.

- *Rate source*: Generates data at the specified number of rows per second, and each output row contains a `timestamp` and `value`. This source is intended for testing purposes. `timestamp` is the `Timestamp` type containing the time of message dispatch. `value` is the `Long` type containing the message count, starting from 0 as the first row.

Let's discuss how to obtain data for processing using File source.

```
// import necessary classes

import org.apache.spark.sql.types.{StructType, StructField, StringType,
IntegerType};

// specify schema

val userSchema = new StructType().add("authorname", "string").
add("publisher", "string")

// create DataFrame using File Source, reads all the .csv files in the data
   directory.

val csvDF = spark.readStream.option("sep", ";").schema(userSchema).csv
("/home/data")

// Create query object to display the contents of the file

val query = csvDF.writeStream.outputMode("append").format("console").start()
```

Note   The csvDF DataFrame is untyped.

Refer to Figure 6-7 for the output.

```
scala> -------------------------------------------
Batch: 0
-------------------------------------------
+----------+---------+
|authorname|publisher|
+----------+---------+
|Subhashini|   Apress|
|   Dharani|   Apress|
+----------+---------+
```

Figure 6-7. *Query result*

Operations on Streaming DataFrames/Datasets

Most of the common operations on DataFrame/Dataset operations are supported for Structured Streaming. Table 6-1 shows student.csv.

Table 6-1. *Student.csv*

S101,John,89
S102,James,78
S103,Jack,90
S104,Joshi,88
S105,Jacob,95

// import required classes

```
import org.apache.spark.sql.functions._
import spark.implicits._
import org.apache.spark.sql.types._;
```

// Specify Schema

```
val studId=StructField("studId",DataTypes.StringType)
val studName=StructField("studName",DataTypes.StringType)
val grade=StructField("grade",DataTypes.IntegerType)
val fields = Array(studId,studName,grade)
val schema = StructType(fields)

case class Student(studId: String, studName: String, grade: Integer)
```

// Create Dataset

```
val csvDS = spark.readStream.option("sep", ",").schema(schema).
csv("/home/data").as[Student]
```

```
// Select the student names where grade is more than 90
```

```
val studNames=csvDS.select("studName").where("grade>90")
```

```
val query = studNames.writeStream.outputMode("append").format("console").
start()
```

The output of this query is shown in Figure 6-8.

```
scala> ---------------------------------------
Batch: 0
---------------------------------------
+--------+
|studName|
+--------+
|   Jacob|
+--------+
```

Figure 6-8. *Student names where grade is more than 90*

We can apply an SQL statement by creating a temporary view as shown here.

```
csvDS.createOrReplaceTempView("student")
```

```
val student=spark.sql("select * from student")
```

```
val query = student.writeStream.outputMode("append").format("console").
start()
```

Refer to Figure 6-9 for the output.

```
scala> ----------------------------------------
Batch: 0
----------------------------------------
+------+--------+-----+
|studId|studName|grade|
+------+--------+-----+
|  S101|    John|   89|
|  S102|   James|   78|
|  S103|    Jack|   90|
|  S104|   Joshi|   88|
|  S105|   Jacob|   95|
+------+--------+-----+
```

Figure 6-9. *Student table*

Let's write a query to find the maximum grade.

```
val gradeMax=spark.sql("select max(grade) from student")
```

```
val query = gradeMax.writeStream.outputMode("complete").format("console").
start()
```

The output for this query is shown in Figure 6-10.

```
scala> ------------------------------------------
Batch: 0
------------------------------------------
+----------+
|max(grade)|
+----------+
|        95|
+----------+
```

Figure 6-10. *Maximum grade*

You can check whether the DataFrame/Dataset has streaming data by issuing the command shown in Figure 6-11.

```
scala> csvDS.isStreaming
res11: Boolean = true
```

Figure 6-11. *isStreaming command*

Note   We need to specify the schema when we perform Structured Streaming from File sources.

Stateful Streaming: Window Operations on Event-Time

Structured Streaming provides straightforward aggregations over a sliding event-time window. This is like grouped aggregation. In a grouped aggregation, aggregate values are maintained for each unique value in the user-specified grouping column. In the same way, in window-based aggregation, aggregate values are maintained for each window into which the event-time of a row falls.

Let us discuss how to count words within 10-minute windows that slide every 5 minutes. For example, count words that are received between 10-minute windows 09:00–09:10, 09:05–09:15, 09:10–09:20, and so on. Suppose the word arrives at 09:07; it should increment the counts in two windows: 09:00–09:10 and 09:05–09:15. Figure 6-12 shows the result tables.

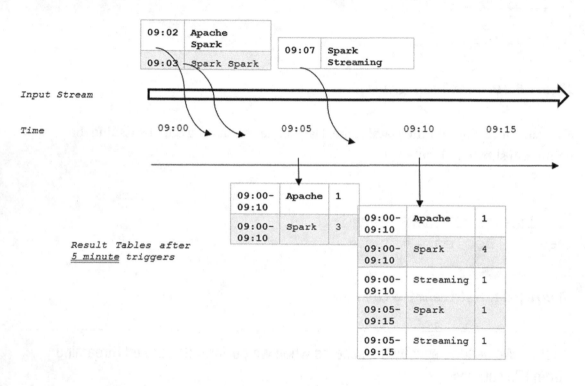

Figure 6-12. *Windowed grouped aggregation with 10-minute windows, sliding every 5 minutes*

```
import java.sql.Timestamp
import org.apache.spark.sql.functions._
import spark.implicits._
```

```
// Create DataFrame representing the stream of input lines from connection
   to host:port

val lines = spark.readStream
.format("socket")
.option("host", "localhost")
.option("port",9999)
.option("includeTimestamp", true)
.load()

// Split the lines into words, retaining timestamps

val words = lines.as[(String, Timestamp)]
.flatMap(line =>line._1.split(" ")
.map(word => (word, line._2)))
.toDF("word", "timestamp")

// Group the data by window and word and compute the count of each group

val windowedCounts = words.groupBy(window($"timestamp", "10 minutes",
"5 minutes"), $"word").count().orderBy("window")

// Start running the query that prints the windowed word counts to the
   console

val query = windowedCounts.writeStream.outputMode("complete").
format("console").option("truncate", "false").start()

query.awaitTermination()
```

The output is shown in Figure 6-13.

```
scala> ----------------------------------------------------
Batch: 0
---------------------------------------------------
+--------------------------------------------------+------+-----+
|window                                            |word  |count|
+--------------------------------------------------+------+-----+
|[2018-04-15 13:00:00.0,2018-04-15 13:10:00.0]|Apache|1    |
|[2018-04-15 13:00:00.0,2018-04-15 13:10:00.0]|Spark |3    |
|[2018-04-15 13:05:00.0,2018-04-15 13:15:00.0]|Spark |3    |
|[2018-04-15 13:05:00.0,2018-04-15 13:15:00.0]|Apache|1    |
+--------------------------------------------------+------+-----+


---------------------------------------------------
Batch: 1
---------------------------------------------------
+--------------------------------------------------+---------+-----+
|window                                            |word     |count|
+--------------------------------------------------+---------+-----+
|[2018-04-15 13:00:00.0,2018-04-15 13:10:00.0]|Streaming|1    |
|[2018-04-15 13:00:00.0,2018-04-15 13:10:00.0]|Apache   |1    |
|[2018-04-15 13:00:00.0,2018-04-15 13:10:00.0]|Spark    |4    |
|[2018-04-15 13:05:00.0,2018-04-15 13:15:00.0]|Streaming|1    |
|[2018-04-15 13:05:00.0,2018-04-15 13:15:00.0]|Spark    |4    |
|[2018-04-15 13:05:00.0,2018-04-15 13:15:00.0]|Apache   |1    |
+--------------------------------------------------+---------+-----+
```

Figure 6-13. *Windowed Structured Streaming output*

Stateful Streaming: Handling Late Data and Watermarking

Structured Streaming maintains the intermediate state for partial aggregates for a long period of time. This helps to update the aggregates of old data correctly when data arrive later than the expected event-time. In short, Spark keeps all the windows forever and waits for the late events forever. Keeping the intermediate state becomes problematic when the volume of data increases. This can be resolved with the help of watermarking. Watermarking allows us to control the state in a bounded way.

Watermarking allows the Spark engine to track the current event-time in the data and clean up the old state accordingly. You can define the watermark of a query by specifying the event-time column and the threshold for how late the data are expected to be in terms of event-time. Late data that arrive within the threshold are aggregated and data that arrive later than the threshold are dropped.

```
val windowedCounts = words
    .withWatermark("timestamp", "10 minutes")
    .groupBy(
        window($"timestamp", "10 minutes", "5 minutes"),
        $"word")
    .count()
```

The following are the conditions for watermarking to clean the aggregation state.

- Output mode should be append or update.

- withWatermark must be called on the same column as the timestamp column used in the aggregate.

- withWatermark must be called before the aggregation for the watermark details to be used.

Triggers

The timing of streaming data processing can be defined with the help of trigger settings, which are described in Table 6-2.

Table 6-2. *Trigger Type*

Trigger Type	Description
Unspecified (default)	The query will be executed in microbatch mode, where microbatches will be generated as soon as the previous microbatch has completed processing.
Fixed interval microbatches	The query will be executed in microbatch mode, where microbatches will be kicked off at the user-specified intervals.
	If the previous microbatch completes within the interval, then the engine will wait until the interval is over before kicking off the next microbatch.
	If the previous microbatch takes longer than the interval to complete (i.e., if an interval boundary is missed), then the next microbatch will start as soon as the previous one completes (i.e., it will not wait for the next interval boundary).
	If no new data are available, then no microbatch will be kicked off.
One-time microbatch	The query will execute only one microbatch to process all the available data and then stop on its own.

Let's discuss how to set the trigger type.

```
import org.apache.spark.sql.streaming.Trigger

// Default trigger (runs microbatch as soon as it can)

df.writeStream
  .format("console")
  .start()

// ProcessingTime trigger with 2-second microbatch interval

df.writeStream
  .format("console")
  .trigger(Trigger.ProcessingTime("5 seconds"))
  .start()
```

```
// One-time trigger

df.writeStream
  .format("console")
  .trigger(Trigger.Once())
  .start()
```

Fault Tolerance

One of the key goals of Structured Streaming is to deliver end-to-end, exactly-once stream processing. To achieve this, Structured Streaming provides streaming sources, an execution engine, and sinks. Every streaming source is assumed to have offsets to track the read position in the stream. The engine uses checkpointing and write-ahead logs to record the offset range of the data that are being processed in each trigger. The streaming sinks are designed to be idempotent for handling reprocessing.

You can specify the checkpoint directory while creating a Spark Session. This code sets a checkpoint directory.

```
import org.apache.spark.sql.SparkSession

val spark: SparkSession = SparkSession.builder
  .master("local[*]")
  .appName("Structured Streaming")
  .config("spark.sql.streaming.checkpointLocation", "/home/checkpoint/")
  .getOrCreate
```

SPARK STRUCTURED STREAMING - EXERCISE 1

1. Write a Spark Structured Streaming application to count the number of WARN messages in a received log stream. Use Netcat to generate the log stream.

2. Extend the code to count WARN messages within 10-minute windows that slide every 5 minutes.

3. Consider the sample employee.csv file shown in Figure 6-14. Create a streaming Dataset to query employee details where project is Spark.

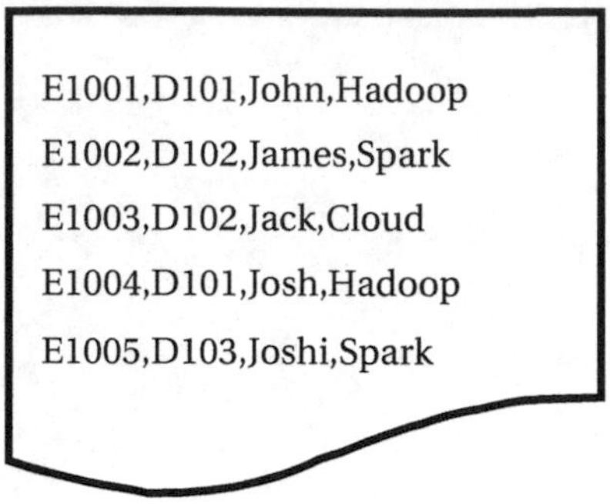

```
E1001,D101,John,Hadoop
E1002,D102,James,Spark
E1003,D102,Jack,Cloud
E1004,D101,Josh,Hadoop
E1005,D103,Joshi,Spark
```

Figure 6-14. employee.csv file

Points to Remember

- Spark Structured Streaming is a fault-tolerant, scalable stream processing engine built on top of Spark SQL.

- In window-based aggregation, aggregate values are maintained for each window into which the event-time of a row falls.

- Watermarking allows the Spark engine to track the current event-time in the data and clean up the old state accordingly.

In next chapter, we will be discussing how to integrate Spark Streaming with Kafka.

Spark Streaming with Kafka

In the previous chapter, you have learned the concepts of Structured Streaming, window-based Structured Streaming, and watermarking. In this chapter, we focus on the basics of Kafka and how to integrate Spark and Kafka.

The recommended background for this chapter is some prior experience with Scala. The mandatory prerequisite for this chapter is completion of the previous chapters assuming that you have practiced all the demos.

We focus on these topics:

- Introduction to Kafka.

- Kafka fundamental concepts.

- Kafka architecture.

- Setting up the Kafka cluster.

- Spark Streaming and Kafka integration.

- Spark Structured Streaming and Kafka integration.

Introduction to Kafka

Apache Kafka is a distributed streaming platform. Apache Kafka is a publishing and subscribing messaging system. It is a horizontally scalable, fault-tolerant system. Kafka is used for these purposes:

- To build real-time streaming pipelines to get data between systems or applications.

- To build real-time streaming applications to transform or react to the streams of data.

© Subhashini Chellappan, Dharanitharan Ganesan 2018
S. Chellappan and D. Ganesan, *Practical Apache Spark*, https://doi.org/10.1007/978-1-4842-3652-9_7

Kafka Core Concepts

- Kafka is run as a cluster on one or more servers.

- The Kafka cluster stores streams of records in categories called *topics*.

- Each record consists of a key, a value, and a timestamp.

Kafka APIs

- *Producer API*: The Producer API enables an application to publish a stream of records to one or more Kafka topics.

- *Consumer API*: The Consumer API enables an application to subscribe to one or more topics and process the stream of records produced to them.

- *Streams API*: The Streams API allows an application to act as a stream processor; that is, this API converts the input streams into output streams.

- *Connector API*: The Connector API allows building and running reusable producers or consumers. These resuable producers or consumers can be used to connect Kafka topics to existing applications or data systems. For example, a connector to a relational database might capture every change to a table.

Figure 7-1 illustrates the Kafka APIs.

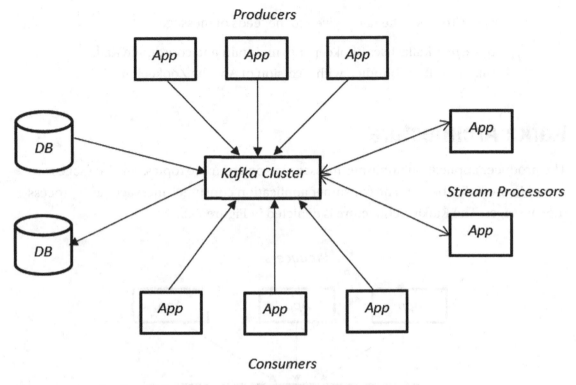

Figure 7-1. *Kafka APIs*

Kafka Fundamental Concepts

Let's cover the fundamental concepts of Kafka.

- *Producer*: The producer is an application that publishes a stream of records to one or more Kafka topics.

- *Consumer*: The consumer is an application that consumes a stream of records from one or more topics and processes the published streams of records.

- *Consumer group*: Consumers label themselves with a consumer group name. One consumer instance within the group will get the message when the message is published to a topic.

- *Broker*: The broker is a server where the published stream of records is stored. A Kafka cluster can contain one or more servers.

- *Topics*: Topics is the name given to the feeds of messages.

- *Zookeeper*: Kafka uses Zookeeper to maintain and coordinate Kafka brokers. Kafka is bundled with a version of Apache Zookeeper.

Kafka Architecture

The producer application publishes messages to one or more topics. The messages are stored in the Kafka broker. The consumer application consumes messages and process the messages. The Kafka architecture is depicted in Figure 7-2.

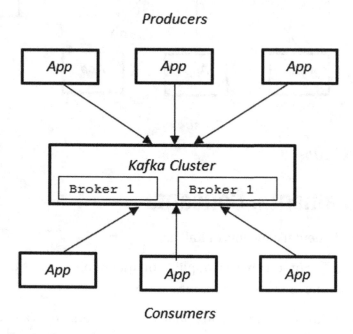

Figure 7-2. *Kafka architecture*

Kafka Topics

We now discuss the core abstraction of Kafka. In Kafka, topics are always multisubscriber entities. A topic can have zero, one, or more consumers. For each topic, a Kafka cluster maintains a partitioned log (see Figure 7-3).

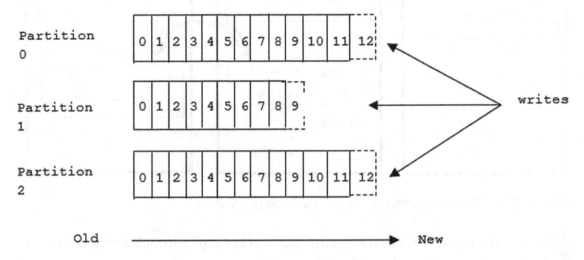

Figure 7-3. Anatonomy of a Kafka topic

The topics are split into multiple partitions. Each parition is an ordered, immutable sequence of records that is continually appended to a structured commit log. The records in the partitions are uniquely identified by sequential numbers called offset. The Kafka cluster persists all the published records for a configurable period whether they are consumed or not. For example, if the retention period is set for two days, the records will be available for two days. After that, they will be discared to free up space. The partitions of the logs are distributed across the server in the Kafka cluster and each partition is replicated across a configurable number of servers to achieve fault tolerance.

Leaders and Replicas

Each partition has one server that acts as the *leader* and zero, one, or more servers that act as *followers*. All the read and write requests for the partition are handled by the leader and followers passively replicate the leader. If the leader fails, any one of the followers becomes the leader automatically. Each server acts as a leader for some of its partitions and a follower for others. This way the load is balanced within the cluster (see Figure 7-4).

179

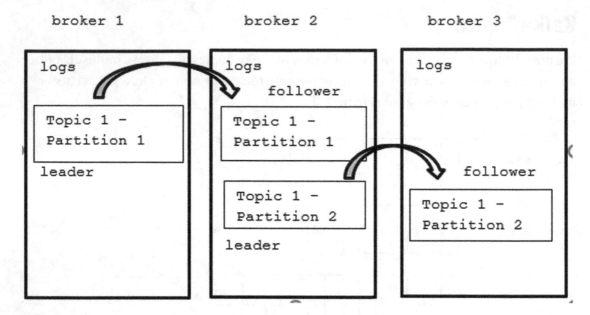

Figure 7-4. *Three brokers, one topic, and two partitions*

When a producer publishes a message to a partition in a topic, first it is forwarded to the leader replica of the partition; the followers then pull the new messages from the leader replica.

The leader commits the message, when enough replicas pull the message. To determine enough replicas, each partition of a topic maintains an in-sync replica set. The in-sync replica (ISR) represents the set of alive replicas that is fully caught up with the leader. Initially, every replica of the partition will be in the ISR. When a new message is published, the leader commits the new message when it reaches all replicas in the ISR. When a follower replica fails, it will be dropped out from the ISR and then the leader commits new messages with remaining replicas.

Setting Up the Kafka Cluster

There are three different ways to set up the Kafka cluster:

- Single node, single broker.

- Single node, multiple broker.

- Multinode, multiple broker.

Let's discuss how to set up a single node, single broker cluster. To do so, follow these steps.

1. Download Kafka from https://kafka.apache.org/downloads.

2. Untar the downloaded Kafka .tgz file.

3. Navigate to the Kafka_2.11-0.11.0.2 folder as shown in Figure 7-5.

```
~/bigdata/kafka_2.11-0.11.0.2$ ▮
```

Figure 7-5. *Kafka folder*

4. Start Zookeeper by issuing the following command.

```
> bin/zookeeper-server-start.sh config/zookeeper.properties
```

5. Open another session, navigate to the Kafka_2.11-0.11.0.2 folder, and start the Kafka broker.

```
> bin/kafka-server-start.sh config/server.properties
```

6. Open another session, navigate to the Kafka_2.11-0.11.0.2 folder, and create a topic named sparkandkafka by issuing following command.

```
> bin/kafka-topics.sh --create --zookeeper localhost:2181
--replication-factor 1 --partitions 1 --topic sparkandkafkatest
```

7. Open another session, navigate to the Kafka_2.11-0.11.0.2 folder, and run the producer. Type a few messages into the console to send to the server (see Figure 7-6).

```
> bin/kafka-console-producer.sh --broker-list localhost:9092
--topic sparkandkafkatest
```

```
>Hello Authors!
>Welcome to Apress Publications!
>▮
```

Figure 7-6. *Publishing messages to the topic sparkandkafkatest*

8. Open another session, navigate to the Kafka_2.11-0.11.0.2 folder, and run the consumer to dump out the messages to standard output (see Figure 7-7).

```
> bin/kafka-console-consumer.sh --bootstrap-server localhost:9092
--topic sparkandkafkatest --from-beginning
```

```
Hello Authors!
Welcome to Apress Publications!
```

Figure 7-7. *Consumer console that dumps the output*

Spark Streaming and Kafka Integration

Let's discuss how to write a Spark application to consume data from the Kafka server that will perform a word count.

1. Download the spark-streaming-kafka-0-8-assembly_2.11-2.1.1.jar file from the following link and place it in the jar folder of Spark.

   http://central.maven.org/maven2/org/apache/spark/spark-streaming-kafka-0-8-assembly_2.11/2.1.1/spark-streaming-kafka-0-8-assembly_2.11-2.1.1.jar

2. Create a build.sbt file as shown in Figure 7-8.

```
name := "spark-Kafka-streaming"
version := "1.0"
scalaVersion := "2.11.8"
libraryDependencies += "org.apache.spark" % "spark-core_2.11" % "2.1.0"
libraryDependencies += "org.apache.spark" % "spark-sql_2.11" % "2.1.0"
libraryDependencies += "org.apache.spark" % "spark-streaming_2.11" % "2.1.0"
libraryDependencies += "org.apache.spark" %% "spark-streaming-kafka-0-8-assembly" % "2.1.1"
```

Figure 7-8. `built.sbt`

3. Create SparkKafkaWordCount.scala as shown here.

```scala
package com.apress.book

import org.apache.spark.sql.{Row, SparkSession}
import org.apache.spark._
import org.apache.spark.streaming._
import org.apache.spark.streaming.kafka._

object SparkKafkaWordCount{

  def main( args:Array[String] ){

    // Create Spark Session and Spark Context

      val spark = SparkSession.builder.appName(getClass.
      getSimpleName).getOrCreate()

    // Get the Spark Context from the Spark Session to create
       Streaming Context
      val sc = spark.sparkContext

    // Create the Streaming Context, interval is 40 seconds

      val ssc = new StreamingContext(sc, Seconds(40))

    // Create Kafka DStream that receives text data from the Kafka
       server.

      val kafkaStream = KafkaUtils.createStream(ssc,
      "localhost:2181","spark-streaming-consumer-group",
      Map("sparkandkafkatest" -> 1))

      val words = kafkaStream.flatMap(x =>  x._2.split(" "))

      val wordCounts = words.map(x => (x, 1)).reduceByKey(_ + _)

    // To print the wordcount result of the stream
      wordCounts.print()
      ssc.start()
      ssc.awaitTermination()
  }
}
```

4. Start the Kafka producer and publish a few messages to a topic
    sparkandkafkatest as shown in Figure 7-9.

    ```
    > bin/kafka-console-producer.sh --broker-list localhost:9092
    --topic sparkandkafkatest
    ```

```
>Hello Authors
>Welcome to Apress Publication
```

Figure 7-9. *Publishing messages to a topic sparkandkafkatest*

5. Build a Spark application using SBT and submit the job to the
    Spark cluster as shown here.

    ```
    spark-submit --class com.apress.book.SparkKafkaWordCount /home/
    data/spark-kafka-streaming_2.11-1.0.jar
    ```

6. The streaming output is shown in Figure 7-10.

```
-------------------------------------------------
Time: 1523876000000 ms
-------------------------------------------------
(Hello,1)
(Authors,1)
(Welcome,1)
(Publication,1)
(Apress,1)
(to,1)
```

Figure 7-10. *Word count output*

Spark Structure Streaming and Kafka Integration

Next we discuss how to integrate Kafka with Spark Structured Streaming.

1.   Start the Spark Shell using this command.

```
> spark-shell --packages 'org.apache.spark:spark-sql-
kafka-0-10_2.11:2.1.0'
```

The package `spark-sql-kafka-0-10_2.11:2.1.0` is required to integrate Spark Structured Streaming and Kafka.

2.   Create a DataFrame to read data from the Kafka server.

```
val readData= spark.readStream.format("kafka").option("kafka.
bootstrap.servers", "localhost:9092").option("subscribe",
"sparkandkafkatest").load()
```

3.   Convert DataFrame into Dataset.

```
val Ds = readData.selectExpr("CAST(key AS STRING)", "CAST( value
AS STRING)").as[(String, String)]
```

4.   Write code to generate the running count of the words as shown here.

```
val wordCounts = Ds.map(_._2.split(" ")).groupBy("value").count()
```

5.   Run a query to print the running count of the word to the console.

```
val query = wordCounts.writeStream.outputMode("complete").
format("console").start()
```

6.   The running count of the word is shown in Figure 7-11.

```
scala> ----------------------------------------------
Batch: 0
----------------------------------------------
+-----+-----+
|value|count|
+-----+-----+
+-----+-----+


----------------------------------------------
Batch: 1
----------------------------------------------
+----------+-----+
|     value|count|
+----------+-----+
|     Hello|    1|
|Subhashini|    1|
+----------+-----+


----------------------------------------------
Batch: 2
----------------------------------------------
+----------+-----+
|     value|count|
+----------+-----+
|     Hello|    2|
|Subhashini|    1|
|   Dharani|    1|
+----------+-----+


----------------------------------------------
Batch: 3
----------------------------------------------
+-----------+-----+
|      value|count|
+-----------+-----+
|Publication|    1|
|      Hello|    2|
| Subhashini|    1|
|     Apress|    1|
|    Dharani|    1|
|    Welcome|    1|
|         to|    1|
+-----------+-----+
```

Figure 7-11. *Running count of the word*

SPARK & KAFKA INTEGRATION - EXERCISE 1

Write a Spark Streaming application to count the number of WARN messages in a received log stream. Use a Kafka producer to generate a log stream.

Points to Remember

- Apache Kafka is a distributed streaming platform. Apache Kafka is a publishing and subscribing messaging system.

- Kafka is run as a cluster on one or more servers.

- The Kafka cluster stores streams of records in categories called topics.

- Each record consists of a key, a value, and a timestamp.

In the next chapter, we discuss the Machine Learning Library of Spark.

CHAPTER 8

Spark Machine Learning Library

In previous chapters, the fundamental components of Spark such as Spark Core, Spark SQL, and Spark Streaming have been covered. In addition to these components, the Spark ecosystem provides an easy way to implement machine learning algorithms through the Spark Machine Learning Library, Spark MLlib. The goal is to implement scalable machine learning easily.

The recommended background for this chapter is to have some prior experience with Scala. Experience with any other programming language is also sufficient. In addition, some familiarity with the command line is beneficial. The mandatory prerequisite for this chapter is to understand the basic concepts of correlation and hypothesis testing. You should also have completed the previous chapters, practiced all the demos, and completed the hands-on exercises given in those chapters.

The examples in the chapter are demostrated using the Scala language.

By end of this chapter, you will be able to do the following:

- Understand the concepts of Spark MLlib.

- Use common learning algorithms such as classification, regression, clustering, and collaborative filtering.

- Construct, evaluate, and tune the machine learning pipelines using Spark MLlib.

Note   It is recommended that you practice the code snippets provided as illustrations and practice the exercises to develop effective knowledge of Spark Machine Learning Libraries.

© Subhashini Chellappan, Dharanitharan Ganesan 2018
S. Chellappan and D. Ganesan, *Practical Apache Spark*, https://doi.org/10.1007/978-1-4842-3652-9_8

What Is Spark MLlib?

Spark MLlib is Spark's collection of machine learning (ML) libraries, which can be used as APIs to implement ML algorithms. The overall goal is to make practical ML scalable and easy. At a high level, Spark MLlib provides tools such as those shown in Figure 8-1.

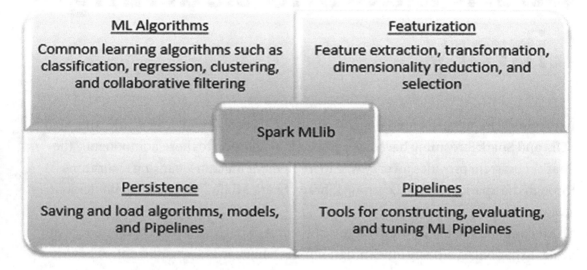

Figure 8-1. *Spark MLlib features*

Spark MLlib APIs

Spark MLlib provides the ML libraries through two different APIs.

1. DataFrame-based API

2. RDD-based API

As of Spark 2.0, the RDD-based APIs in the `spark.mllib` package have been taken back for maintenance and are not deprecated. Now the primary API for ML is the DataFrame-based API in the `spark.ml` package.

However, MLlib still supports the RDD-based API in the `spark.mllib` package with some bug fixes. Spark MLlib will not add any new features to the RDD-based API, however. Also, the RDD-based API is expected to be removed from MLlib in Spark 3.0.

Why are DataFrame-based APIs better than RDD-based APIs? Here are three reasons (see Figure 8-2).

1. DataFrames provide a more user-friendly API than RDDs. The many benefits of DataFrames include Spark data sources, SQL/DataFrame queries, and uniform APIs across languages.

2. The DataFrame-based API for MLlib provides a uniform API across ML algorithms and across multiple languages.

3. DataFrames facilitate practical ML pipelines, particularly feature transformations.

Spark MLlib	Basic Statistics
Dataframe based APIs	Pipelines
	Extracting, transforming and selecting features
	•Regression and Classification
	•Clustering
	•Collaborative filtering
	•Frequent Pattern Mining
	•Model selection and ML tuning

Figure 8-2. *Spark MLlib DataFrame-based API features*

Note   Before we start with basic statistics, it is higly recommended that you understand vectors and the importance of sparse vectors and dense vectors. Later in this chapter, we explain the concept of vectors with a simple example in Scala.

Vectors in Scala

A vector is an immutable collection in Scala. Although it is immutable, a vector can be added to and updated. The operator :+ is used to add any elements to the end of a vector and the operator +: is used to add the element to the start of a vector.
Let's start by creating the empty vector using

```
scala.collection.immutable.Vector.empty
```

and add the elements to the start and the end of the vector.

```
val v1 = scala.collection.immutable.Vector.empty
println(v1)

val v2 = v1 :+ 5
println(v2)

val v3 = v2 :+ 10 :+ 20
println(v3)
```

The output is shown in Figure 8-3.

```
scala> val v1 = scala.collection.immutable.Vector.empty
v1: scala.collection.immutable.Vector[Nothing] = Vector()

scala> println(v1)
Vector()

scala> val v2 = v1 :+ 5
v2: scala.collection.immutable.Vector[Int] = Vector(5)

scala> println(v2)
Vector(5)

scala> val v3 = v2 :+ 10 :+ 20
v3: scala.collection.immutable.Vector[Int] = Vector(5, 10, 20)

scala> println(v3)
Vector(5, 10, 20)
```

Figure 8-3.   *Vectors in Scala*

The vector values can be changed using the updated() method based on the index of elements.

```
val v3_changed = v3.updated(2,100)
println(v3_changed)
```

The output is shown in Figure 8-4.

```
scala> val v3_changed = v3.updated(2,100)
v3_changed: scala.collection.immutable.Vector[Int] = Vector(5, 10, 100)

scala> println(v3_changed)
Vector(5, 10, 100)
```

Figure 8-4.   *Updating vectors in Scala*

Vector Representation in Spark

The vectors can be defined as dense vectors or sparse vectors. For example, let's say we want to create the following vector: {0, 2, 0, 4}.

Because the implementation of vectors in a programming language occurs as a one-dimensional array of elements, the vector is said to be sparse if many elements have zero values. From a storage perspective, it is not good to store the zero values or null values. It is better to represent the vector as a sparse vector by specifying the location of nonzero values only. The sparse vector is represented as

```
sparse(int size, int[] indices, double[] values)
```

This method creates a sparse vector where the first argument is size, the second argument is the indexes where a value exists, and the last argument is the values on these indexes. The other elements of this vector have values of zero.

The Vector class of org.apache.spark.mllib.linalg has multiple methods to create the dense and sparse vectors. First, start the Spark Shell (see Figure 8-5).

```
Spark context Web UI available at http://10.0.2.15:4040
Spark context available as 'sc' (master = local[*], app id = local-1522988807355).
Spark session available as 'spark'.
Welcome to

    ____              __
   / __/__  ___ _____/ /__
  _\ \/ _ \/ _ `/ __/  '_/
 /___/ .__/\_,_/_/ /_/\_\   version 2.2.1
    /_/

Using Scala version 2.11.8 (Java HotSpot(TM) 64-Bit Server VM, Java 1.8.0_77)
Type in expressions to have them evaluated.
Type :help for more information.

scala>
```

Figure 8-5.   *Starting Spark Shell*

Next, create the dense vector by importing Vectors from the spark.ml package.

```
import org.apache.spark.ml.linalg.Vectors
val denseVector=vectors.dense(1,2,0,0,5)
print(denseVector)
```

The output is shown in Figure 8-6.

```
scala> import org.apache.spark.ml.linalg.Vectors
import org.apache.spark.ml.linalg.Vectors

scala> val denseVector=Vectors.dense(1,2,0,0,5)
denseVector: org.apache.spark.ml.linalg.Vector = [1.0,2.0,0.0,0.0,5.0]

scala> print(denseVector)
[1.0,2.0,0.0,0.0,5.0]
```

Figure 8-6. *Dense vectors in Spark*

The same can be created as sparse vectors by specifying the size and indices of nonzero elements.

As discussed earlier, the sparse vector is represented as

```
Vectors.sparse(size, indices, values)
```

where indices are represented as an integer array and values as a double array.

```
Val sparseVector = Vectors.sparse(5,Array(0,1,4),Array(1.0,2.0,5.0))
print(sparseVector)
```

The output is shown in Figure 8-7.

```
scala> val sparseVector=Vectors.sparse(5,Array(0,1,4),Array(1.0,2.0,5.0))
sparseVector: org.apache.spark.ml.linalg.Vector = (5,[0,1,4],[1.0,2.0,5.0])

scala> print(sparseVector)
(5,[0,1,4],[1.0,2.0,5.0])
```

Figure 8-7. *Sparse vectors in Spark*

In the preceding example, the sparse vector is created with size 5. The nonzero elements are represented in the indices [0,1,4] and the values are [1.0,2.0,5.0], respectively.

Note   It is mandatory to specify the values in the sparse vector as a double array.

Basic Statistics

The most important statistical components of Spark MLlib are correlation and hypothesis testing.

Correlation

The basic operation in the statistics is calculating the correlation between the two series of data. The `spark.ml` package provides the flexiblity to calculate the pairwise correlation among many series of data.

There are two currently supported correlation methods in the `spark.ml` package: Pearson correlation and Spearman correlation. Correlation computes the correlation matrix for input data set of vectors using the specified method of correlation.

The *Pearson correlation* is a number between –1 and 1 that indicates the extent to which two variables are linearly related. The Pearson correlation is also known as the product–moment correlation coefficient (PMCC) or simply correlation.

The *Spearman rank-order correlation* is the nonparametric version of the Pearson product–moment correlation. The Spearman correlation coefficient measures the strength and direction of association between two ranked variables.

The output will be the DataFrame that contains the correlation matrix of the column of vectors.

Import the following classes:

```
import org.apache.spark.ml.linalg.Matrix
import org.apache.spark.ml.linalg.Vectors
import org.apache.spark.ml.stat.Correlation
import org.apache.spark.sql.Row
```

Then create a sample Dataset of vectors:

```
val data = List(
  Vectors.sparse(4, Array(0,3), Array(1.0, -2.0)),
  Vectors.dense(4.0, 5.0, 0.0, 3.0),
  Vectors.dense(6.0, 7.0, 0.0, 8.0),
  Vectors.sparse(4, Array(0,3), Array(9.0, 1.0))
)
```

Create a DataFrame using Spark SQL's `toDF()` method:

```
val dataFrame = sampleData.map(Tuple1.apply).toDF("features")
```

Create the correlation matrix by passing the DataFrame to the `Correlation.corr()` method.

```
val Row(coeff: Matrix) = Correlation.corr(dataFrame,"features").head
println(s"The Pearson correlation matrix:\n\n$coeff")
```

Figure 8-8 shows the execution steps in Spark Shell.

```
scala> import org.apache.spark.ml.linalg.Matrix
import org.apache.spark.ml.linalg.Matrix

scala> import org.apache.spark.ml.linalg.Vectors
import org.apache.spark.ml.linalg.Vectors

scala> import org.apache.spark.ml.stat.Correlation
import org.apache.spark.ml.stat.Correlation

scala> import org.apache.spark.sql.Row
import org.apache.spark.sql.Row

scala>

scala> val data = List(
     |     Vectors.sparse(4, Array(0,3), Array(1.0, -2.0)),
     |     Vectors.dense(4.0, 5.0, 0.0, 3.0),
     |     Vectors.dense(6.0, 7.0, 0.0, 8.0),
     |     Vectors.sparse(4, Array(0,3), Array(9.0, 1.0))
     | )
data: List[org.apache.spark.ml.linalg.Vector]

scala>

scala> val dataFrame = sampleData.map(Tuple1.apply).toDF("features")
dataFrame: org.apache.spark.sql.DataFrame = [features: vector]

scala> val Row(coeff: Matrix) = Correlation.corr(dataFrame,"features").head
18/04/06 06:35:49 WARN PearsonCorrelation: Pearson correlation matrix contains NaN values
coeff: org.apache.spark.ml.linalg.Matrix =
1.0                   0.055641488407465814   NaN   0.4004714203168137
0.055641488407465814  1.0                    NaN   0.9135958615342522
NaN                   NaN                    1.0   NaN
0.4004714203168137    0.9135958615342522     NaN   1.0

scala> println(s"The Pearson correlation matrix:\n\n$coeff")
The Pearson correlation matrix:

1.0                   0.055641488407465814   NaN   0.4004714203168137
0.055641488407465814  1.0                    NaN   0.9135958615342522
NaN                   NaN                    1.0   NaN
0.4004714203168137    0.9135958615342522     NaN   1.0
```

Figure 8-8. *Pearson correlation matrix calculation in Spark*

The complete code excerpt for correlation matrix formation is given here.

```
package com.apress.statistics

import org.apache.spark.ml.linalg.Matrix
import org.apache.spark.ml.linalg.Vectors
import org.apache.spark.ml.stat.Correlation
```

```scala
import org.apache.spark.sql.Row
import org.apache.spark.sql.SparkSession

object PearsonCorrelationDemo {

  def main(args: Array[String]): Unit = {

    val sparkSession = SparkSession.builder
                        .appName("ApressCorrelationExample")
                        .master("local[*]")
                        .getOrCreate()

    import sparkSession.implicits._

    val sampleData = List(
      Vectors.sparse(4, Array(0, 3), Array(1.0, -2.0)),
      Vectors.dense(4.0, 5.0, 0.0, 3.0),
      Vectors.dense(6.0, 7.0, 0.0, 8.0),
      Vectors.sparse(4, Array(0, 3), Array(9.0, 1.0)))

    val dataFrame = sampleData.map(Tuple1.apply).toDF("features")

    val Row(coeff: Matrix) = Correlation.corr(dataFrame,"features").head

    println(s"The Pearson correlation matrix:\n $coeff")

    sparkSession.stop()

  }
}
```

Note   To execute the given code in any integrated development environment (IDE) that supports Scala, it is mandatory to add the Scala library to the project workspace and all the Spark jars to the classpath.

The Spearman correlation matrix can be calculated by specifying the type in

```scala
val Row(coeff: Matrix) = Correlation.corr(df, "features", "spearman").head
```

The calculation is displayed in Figure 8-9.

```
scala> val Row(coeff: Matrix) = Correlation.corr(dataFrame, "features", "spearman").head

coeff: org.apache.spark.ml.linalg.Matrix =
1.0                   0.10540925533894532  NaN  0.40000000000000174
0.10540925533894532   1.0                  NaN  0.9486832980505141
NaN                   NaN                  1.0  NaN
0.40000000000000174   0.9486832980505141   NaN  1.0

scala> println(s"Spearman correlation matrix:\n$coeff")
Spearman correlation matrix:
1.0                   0.10540925533894532  NaN  0.40000000000000174
0.10540925533894532   1.0                  NaN  0.9486832980505141
NaN                   NaN                  1.0  NaN
0.40000000000000174   0.9486832980505141   NaN  1.0
```

Figure 8-9. *Spearman correlation matrix calculation in Spark*

Hypothesis Testing

Hypothesis testing is conducted to determine whether the result is statistically significant or not. Currently the spark.ml package supports the Pearson chi-square (χ^2) tests for independence.

ChiSquareTest conducts the Pearson independence test for each feature against the label. For each feature, the (feature, label) pairs are converted into a contingency matrix for which the chi-square statistic is computed.

Import the following ChiSquareTest class from the spark.ml package:

```
import org.apache.spark.ml.linalg.Vector
import org.apache.spark.ml.linalg.Vector
import org.apache.spark.ml.stat.ChiSquareTest
```

The ChiSquareTest can be conducted on the DataFrame by this method.

```
ChiSquareTest.test(dataFrame, "features", "label").head
```

Figure 8-10 shows the execution steps for ChiSquareTest in Spark Shell.

```
scala> import org.apache.spark.ml.linalg.Vector
import org.apache.spark.ml.linalg.Vector

scala> import org.apache.spark.ml.linalg.Vectors
import org.apache.spark.ml.linalg.Vectors

scala> import org.apache.spark.ml.stat.ChiSquareTest
import org.apache.spark.ml.stat.ChiSquareTest

scala>

scala> val data = List(
     |     (0.0, Vectors.dense(0.5, 15.0)),
     |     (0.0, Vectors.dense(1.5, 20.0)),
     |     (1.0, Vectors.dense(1.5, 35.0)),
     |     (0.0, Vectors.dense(3.5, 35.0)),
     |     (0.0, Vectors.dense(3.5, 45.0)),
     |     (1.0, Vectors.dense(3.5, 55.0))
     | )
data: List[(Double, org.apache.spark.ml.linalg.Vector)]

scala>

scala> val dataFrame = data.toDF("label", "features")
dataFrame: org.apache.spark.sql.DataFrame = [label: double, features: vector]

scala> val test = ChiSquareTest.test(dataFrame, "features", "label").head
test: org.apache.spark.sql.Row

scala> println(s"pValues = ${test.getAs[Vector](0)}")
pValues = [0.6872892787909721,0.44089552967916945]

scala> println(s"degreesOfFreedom ${test.getSeq[Int](1).mkString("[", "," ,"]")}")
degreesOfFreedom [2,4]

scala> println(s"statistics ${test.getAs[Vector](2)}")
statistics [0.75,3.7500000000000004]
```

Figure 8-10. *Hypothesis testing: Chi-square test*

The complete code snippet for hypothesis testing with ChiSquareTest (using the spark. ml package) is given here.

```
package com.apress.statistics

import org.apache.spark.ml.linalg.Vector
import org.apache.spark.ml.linalg.Vectors
import org.apache.spark.ml.stat.ChiSquareTest
import org.apache.spark.sql.SparkSession

object HypothesisTestingExample {

  def main(args: Array[String]): Unit = {
```

```scala
    val sparkSession = SparkSession.builder
                        .appName("ApressHypothesisExample")
                        .master("local[*]")
                        .getOrCreate()

    import sparkSession.implicits._

    val sampleData = List(
      (0.0, Vectors.dense(0.5, 15.0)),
      (0.0, Vectors.dense(1.5, 20.0)),
      (1.0, Vectors.dense(1.5, 35.0)),
      (0.0, Vectors.dense(3.5, 35.0)),
      (0.0, Vectors.dense(3.5, 45.0)),
      (1.0, Vectors.dense(3.5, 55.0)))

    val dataFrame = sampleData.toDF("label", "features")
    val test = ChiSquareTest.test(dataFrame, "features", "label").head

  println(s"pValues = ${test.getAs[Vector](0)}")
  println(s"degreesOfFreedom ${test.getSeq[Int](1).mkString("[",",","]")}")
  println(s"statistics ${test.getAs[Vector](2)}")

  }
}
```

Note   To execute the given code in any IDE that supports Scala, it is mandatory to add the Scala library to the project workspace and all the Spark jars to the classpath.

Extracting, Transforming, and Selecting Features

Extraction deals with extracting the features with the raw data. Transformation deals with scaling, converting, and modifying the features extracted from the raw data. Selection deals with taking a sample or subset from large set of features.

Figure 8-11 explains the list of the available and most commonly used feature extractors, feature transformers, and feature selectors.

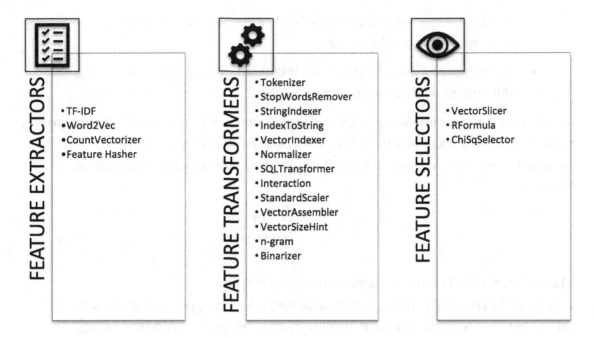

FEATURE EXTRACTORS
- TF-IDF
- Word2Vec
- CountVectorizer
- Feature Hasher

FEATURE TRANSFORMERS
- Tokenizer
- StopWordsRemover
- StringIndexer
- IndexToString
- VectorIndexer
- Normalizer
- SQLTransformer
- Interaction
- StandardScaler
- VectorAssembler
- VectorSizeHint
- n-gram
- Binarizer

FEATURE SELECTORS
- VectorSlicer
- RFormula
- ChiSqSelector

Figure 8-11. *Feature extractors, transformers, and selectors*

Note   Refer to the Spark Machine Learning Library module in the Apache Spark documentation for the complete list of feature extractors, feature transformers, and feature selectors.

Feature Extractors

Feature extraction is the process of transforming the input data into a set of features that can represent the input data very well. The various available feature extractors in Spark MLlib are explained later in this chapter.

Term Frequency–Inverse Document Frequency (TF–IDF)

TF–IDF is a vectorization method to understand the importance of a term to the document in the corpus. The notations are given here.

```
Term - t, Document - d,    Corpus - D
```

- *Term frequency TF(t,d):* This is defined as the number of times the term appears in the document.

- *Document frequency DF(t,D):* This is defined as the number of documents containing the term.

If a term appears very frequently in the corpus, it won't carry any special information about a document. Examples include *a, is, are,* and *for.* It is very easy to overemphasize these terms because they appear very often and carry little information about the document.

$$IDF(t,D) = \log \frac{|D|+1}{DF(t,D)+1}$$

where |D| is the total number of documents in the corpus.

This logarithm is used to make the IDF value zero if a term appears in all documents. A smoothing term is applied to avoid dividing by zero for terms outside the corpus.

- *TF–IDF, Term Frequency Inverse Document Frequency:* This is the product of term frequency and inverse document frequency.

$$TFIDF(t,d,D) = TF(t,d) * IDF(t,D)$$

- *Term frequency generation:* The `HashingTF` and `CountVectorizer` can be used to generate the term frequency vectors. `HashingTF` is a transformer that generates fixed-length feature vectors from the input set of terms. `CountVectorizer` creates the vector of term counts from text documents.

- *Inverse document frequency generation:* IDF is an estimator that fits on a data set and produces an `IDFModel`. The `IDFModel` scales the features created from the `HashingTF` or `CountVectorizer` by down-weighting the frequently appearing features in the corpus.

Example

Execute the following example in the shell and observe the output from each step (see Figure 8-12).

```
import org.apache.spark.ml.feature.HashingTF

import org.apache.spark.ml.feature.IDF

import org.apache.spark.ml.feature.Tokenizer

val rawData = spark.createDataFrame(Seq(
                            (0.0, "This is spark book"),
                            (0.0, "published by Apress publications"),
                            (1.0, "Dharanitharan wrote this book"))).
                            toDF("label", "sentence")

val tokenizer = new Tokenizer().setInputCol("sentence").
setOutputCol("words")

val wordsData = tokenizer.transform(rawData)

val hashingTF = new HashingTF().setInputCol("words")
                            .setOutputCol("rawFeatures")
                            .setNumFeatures(20)

val featurizedData = hashingTF.transform(wordsData)

val idf = new IDF().setInputCol("rawFeatures").setOutputCol("features")

val idfModel = idf.fit(featurizedData)

val rescaledData = idfModel.transform(featurizedData)

rescaledData.select("label", "features").show(false)
```

```
scala> import org.apache.spark.ml.feature.HashingTF
import org.apache.spark.ml.feature.HashingTF

scala> import org.apache.spark.ml.feature.IDF
import org.apache.spark.ml.feature.IDF

scala> import org.apache.spark.ml.feature.Tokenizer
import org.apache.spark.ml.feature.Tokenizer

scala> val rawData = spark.createDataFrame(Seq(
     |    (0.0, "This is spark book"),
     |    (0.0, "published by Apress publications"),
     |    (1.0, "Dharanitharan wrote this book")
     | )).toDF("label", "sentence")
rawData: org.apache.spark.sql.DataFrame

scala> val tokenizer = new Tokenizer().
     |  setInputCol("sentence").
     |  setOutputCol("words")
tokenizer: org.apache.spark.ml.feature.Tokenizer

scala> val wordsData = tokenizer.transform(rawData)
wordsData: org.apache.spark.sql.DataFrame

scala> val hashingTF = new HashingTF().
     |   setInputCol("words").
     |   setOutputCol("rawFeatures").
     |   setNumFeatures(10)
hashingTF: org.apache.spark.ml.feature.HashingTF

scala> val featurizedData = hashingTF.transform(wordsData)
featurizedData: org.apache.spark.sql.DataFrame

scala> val idf = new IDF().setInputCol("rawFeatures").
     |   setOutputCol("features")
idf: org.apache.spark.ml.feature.IDF = idf_3da066d50a3d

scala> val idfModel = idf.fit(featurizedData)
idfModel: org.apache.spark.ml.feature.IDFModel

scala> val rescaledData = idfModel.
     |   transform(featurizedData)
rescaledData: org.apache.spark.sql.DataFrame


scala> rescaledData.select("label", "features").show(false)
+-----+----------------------------------------------------------------+
|label|features                                                        |
+-----+----------------------------------------------------------------+
|0.0  |(10,[1,3,5],[0.28768207245178085,0.0,0.6931471805599453])|
|0.0  |(10,[3,7],[0.0,0.6931471805599453])                       |
|1.0  |(10,[0,1,3],[0.6931471805599453,0.28768207245178085,0.0])|
+-----+----------------------------------------------------------------+
```

Figure 8-12. *TF–IDF ➤ HashingTF term frequency extractor*

The CountVectorizer can also be used for creating the feature vectors as shown here (see Figure 8-13).

```
import org.apache.spark.ml.feature.CountVectorizer

val rawData = spark.createDataFrame(Seq(
                                (0.0, "This is spark book"),
                                (0.0, "published by Apress publications"),
                                (1.0, "Dharanitharan wrote this book"))).
                                toDF("label", "sentence")

val couvtVecModel = new CountVectorizer()
                    .setInputCol("sentence").setOutputCol("features")
                    .setVocabSize(3).setMinDF(2).fit(rawData)

couvtVecModel.transform(rawData).show(false)
```

```
scala> import org.apache.spark.ml.feature.CountVectorizer
import org.apache.spark.ml.feature.CountVectorizer

scala>

scala> val rawData = spark.createDataFrame(Seq(
     |  (0.0, Array("This","is","spark", "book")),
     |  (0.0, Array("published", "by", "Apress", "publications")),
     |  (1.0, Array("Dharanitharan", "wrote", "this", "book")))).
     |  toDF("label", "sentence")
rawData: org.apache.spark.sql.DataFrame

scala> val couvtVecModel = new CountVectorizer().
     |  setInputCol("sentence").
     |  setOutputCol("features").
     |  setVocabSize(3).
     |  setMinDF(2).fit(rawData)
couvtVecModel: org.apache.spark.ml.feature.CountVectorizerModel

scala> couvtVecModel.transform(rawData).show(false)
+-----+------------------------------------+-------------+
|label|sentence                            |features     |
+-----+------------------------------------+-------------+
|0.0  |[This, is, spark, book]             |(1,[0],[1.0])|
|0.0  |[published, by, Apress, publications]|(1,[],[])   |
|1.0  |[Dharanitharan, wrote, this, book]  |(1,[0],[1.0])|
+-----+------------------------------------+-------------+
```

Figure 8-13. *TF–IDF* ➤ *CountVectorizer term frequency extractor*

Feature Transformers

The transformers implement a method `transform()`, which converts one DataFrame into another, generally by appending or removing one or more columns. The various available feature transformers in Spark MLlib are explained later in this chapter.

Tokenizer

The process of splitting a full sentence into individual words is called *tokenization*. Figure 8-14 shows the process of splitting sentences into sequences of words using the Tokenizer.

```scala
scala> import org.apache.spark.ml.feature.Tokenizer
import org.apache.spark.ml.feature.Tokenizer

scala> import org.apache.spark.sql.functions._
import org.apache.spark.sql.functions._

scala> val rawData = spark.createDataFrame(Seq(
     |    (0.0, "This is spark book"),
     |    (0.0, "published by Apress publications"),
     |    (1.0, "Dharanitharan wrote this book")
     |    )).toDF("label", "sentence")
rawData: org.apache.spark.sql.DataFrame = [label: double, sentence: string]

scala> val tokenizer = new Tokenizer().
     |   setInputCol("sentence").
     |   setOutputCol("words")
tokenizer: org.apache.spark.ml.feature.Tokenizer = tok_4ae4da6d943f

scala> val countTokens = udf { (words: Seq[String]) => words.length }
countTokens: org.apache.spark.sql.expressions.UserDefinedFunction

scala> val tokenized = tokenizer.transform(rawData)
tokenized: org.apache.spark.sql.DataFrame

scala> tokenized.select("sentence", "words").
     |   withColumn("tokens", countTokens(col("words"))).
     |   show(false)
+--------------------------------+-----------------------------------------+------+
|sentence                        |words                                    |tokens|
+--------------------------------+-----------------------------------------+------+
|This is spark book              |[this, is, spark, book]                  |4     |
|published by Apress publications|[published, by, apress, publications]    |4     |
|Dharanitharan wrote this book   |[dharanitharan, wrote, this, book]       |4     |
+--------------------------------+-----------------------------------------+------+
```

Figure 8-14. *Tokenization using the Tokenizer transformer*

StopWordsRemover

The StopWordsRemover transformer (see Figure 8-15) is used to exclude the set of words that does not carry much meaning from the input. For example, *I*, *was*, *is*, *an*, *the*, and *for* can be the stop words, because they do not carry much meaning in the sentence to create the features.

```
scala> import org.apache.spark.ml.feature.StopWordsRemover
import org.apache.spark.ml.feature.StopWordsRemover

scala>

scala> val wordsRemover = new StopWordsRemover().
     | setInputCol("rawData").
     | setOutputCol("filtered")
wordsRemover: org.apache.spark.ml.feature.StopWordsRemover = stopWords_a2c7ec9

scala>

scala> val input = spark.createDataFrame(Seq(
     | (0, Seq("This","is","spark", "book")),
     | (1, Seq("published","by", "Apress", "publications")))).
     | toDF("id", "rawData")
input: org.apache.spark.sql.DataFrame = [id: int, rawData: array<string>]

scala> wordsRemover.transform(input).show(false)
+---+-----------------------------------------+------------------------------------+
|id |rawData                                  |filtered                            |
+---+-----------------------------------------+------------------------------------+
|0  |[This, is, spark, book]                  |[spark, book]                       |
|1  |[published, by, Apress, publications]    |[published, Apress, publications]   |
+---+-----------------------------------------+------------------------------------+
```

Figure 8-15. *StopWordsRemover transformer*

The input to StopWordsRemover is sequence of strings (i.e., the output of Tokenizer) and it filters all the stop words specified in the stopWords parameter.

StopWordsRemover.loadDefaultStopWords(language) provides the default stop words in any language. For example, the default language is English.

Also, the custom stop words can be specified using the stopWords parameter as shown here (see Figure 8-16).

```
val wordsRemover = new StopWordsRemover().
                   setInputCol("rawData").
                   setOutputCol("filtered").
                   setStopWords(Array("book","apress"))
```

```
scala> import org.apache.spark.ml.feature.StopWordsRemover
import org.apache.spark.ml.feature.StopWordsRemover

scala>

scala> val wordsRemover = new StopWordsRemover().
     | setInputCol("rawData").
     | setOutputCol("filtered").
     | setStopWords(Array("book","apress"))
wordsRemover: org.apache.spark.ml.feature.StopWordsRemover = stopWords_e473

scala>

scala> val input = spark.createDataFrame(Seq(
     | (0, Seq("This","is","spark","BOOK", "book")),
     | (1, Seq("published","by", "Apress", "publications")))).
     | toDF("id", "rawData")
input: org.apache.spark.sql.DataFrame = [id: int, rawData: array<string>]

scala>

scala> wordsRemover.transform(input).show(false)
+---+------------------------------------------+-------------------------------+
|id |rawData                                   |filtered                       |
+---+------------------------------------------+-------------------------------+
|0  |[This, is, spark, BOOK, book]             |[This, is, spark]              |
|1  |[published, by, Apress, publications]     |[published, by, publications]  |
+---+------------------------------------------+-------------------------------+
```

Figure 8-16. *StopWordsRemover transformer with* stopWords *parameter*

By default, the caseSensitive parameter is false. Hence, it removes the specified stop words irrespective of case. It can be changed by specifying the caseSensitive parameter as shown in Figure 8-17.

```
scala> val wordsRemover = new StopWordsRemover().
     | setInputCol("rawData").
     | setOutputCol("filtered").
     | setStopWords(Array("book","apress")).
     | setCaseSensitive(false)
wordsRemover: org.apache.spark.ml.feature.StopWordsRemover = stopWords_990

scala>

scala> wordsRemover.transform(input).show(false)
+---+------------------------------------------+-------------------------------+
|id |rawData                                   |filtered                       |
+---+------------------------------------------+-------------------------------+
|0  |[This, is, spark, BOOK, book]             |[This, is, spark]              |
|1  |[published, by, Apress, publications]     |[published, by, publications]  |
+---+------------------------------------------+-------------------------------+
```

Figure 8-17. *StopWordsRemover transformer with* caseSensitive *parameter*

Figure 8-18 illustrates the flow of the Tokenizer and StopWords transformers.

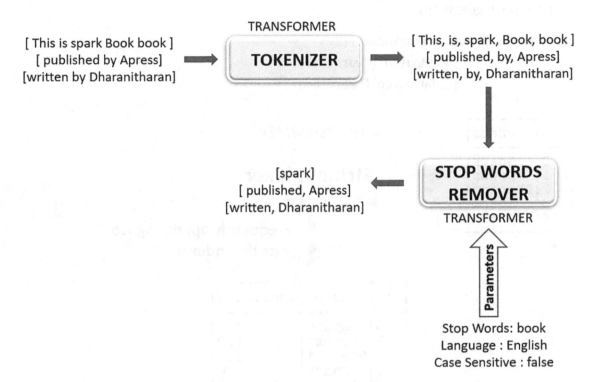

Figure 8-18. *Feature transformers illustration: Tokenizer and StopWords transformers*

StringIndexer

The StringIndexer encodes the labels of a string column to a column of label indices. The indices are in [0, numLabels), ordered by label frequencies, so the most frequent label gets index 0. For example:

```
val input = spark.createDataFrame(Seq(
          (0, "Spark"),(1, "Apress"),(2, "Dharani"),(3, "Spark"),
          (4,"Apress"))).toDF("id", "words")
```

This line creates a DataFrame with columns id and words. Words is a string column with three labels: "Spark", "Apress", and "Dharani".

Applying StringIndexer with words as the input column and the wordIndex as the output column (see Figure 8-19):

```
val indexer = new StringIndexer().
              setInputCol("words").
              setOutputCol("wordIndex")
```

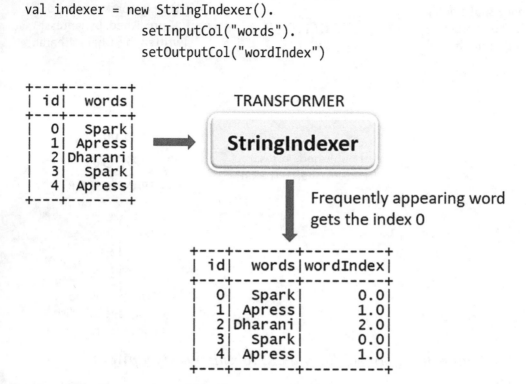

Figure 8-19. *StringIndexer transformer*

The word "Spark" gets index 0 because it is the most frequent, followed by "Apress" with index 1 and "Dharani" with index 2 (see Figure 8-20).

When the downstream pipeline components like Estimator or any Transformer uses this string-indexed label, it is must set the input column of the respective component to this string-indexed column name. Generally, the input column is set by the setInputCol property.

```scala
scala> import org.apache.spark.ml.feature.StringIndexer
import org.apache.spark.ml.feature.StringIndexer

scala>

scala> val input = spark.createDataFrame(Seq(
     |  (0, "Spark"),
     |  (1, "Apress"),
     |  (2, "Dharani"),
     |  (3, "Spark"),
     |  (4, "Apress"))).
     |  toDF("id", "words")
input: org.apache.spark.sql.DataFrame = [id: int, words: string]

scala>

scala> val indexer = new StringIndexer().
     |  setInputCol("words").
     |  setOutputCol("wordIndex")
indexer: org.apache.spark.ml.feature.StringIndexer = strIdx_8260d7

scala>

scala> val indexed = indexer.fit(input).transform(input)
indexed: org.apache.spark.sql.DataFrame = [id: int, words: string

scala>

scala> indexed.show()
+---+-------+---------+
| id|  words|wordIndex|
+---+-------+---------+
|  0|  Spark|      0.0|
|  1| Apress|      1.0|
|  2|Dharani|      2.0|
|  3|  Spark|      0.0|
|  4| Apress|      1.0|
+---+-------+---------+
```

Figure 8-20.   *Feature transformer: StringIndexer*

Feature Selectors

The feature selectors are used to select the required features based on indices. The available feature selectors in Spark MLlib are explained later in this chapter.

VectorSlicer

VectorSlicer takes the feature vector as input and outputs a new feature vector with a subset of the original features. It is useful for extracting required features from a vector column.

VectorSlicer accepts a vector column with specified indices, then outputs a new vector column with values that are selected through the indices. There are two types of indices.

- *Integer indices*: This represents the indices into the vector. It is represented by setIndices().

- *String indices*: This represents the names of features into the vector and represented by setNames(). This requires the vector column to have an AttributeGroup because the implementation matches on the name field of an Attribute.

Create a DataFrame with feature vectors and map the attributes using Attribute groups.

```
val data = Arrays.asList(
                    Row(Vectors.dense(2.5, 2.9, 3.0)),
                    Row(Vectors.dense(-2.0, 2.3, 0.0))
                    )
val defaultAttr = NumericAttribute.defaultAttr

val attrs = Array("col1", "col2", "col3").map(defaultAttr.withName)

val attrGroup = new AttributeGroup(
                            "InFeatures",
                            attrs.asInstanceOf[Array[Attribute]]
                            )
```

```
val dataset = spark.createDataFrame(
                              data,
                              StructType(Array(attrGroup.
                              toStructField()))
                              )
```

Then create a VectorSlicer,

```
val slicer = new VectorSlicer()
                .setInputCol("InFeatures")
                .setOutputCol("SelectedFeatures")
```

Set the index to slicer to select the feature that is required. For example, if col1 is required, set the index as 0 or name as "col1".

```
slicer.setIndices(Array(0))
```

--or--

```
slicer.setNames(Array("col1"))
```

Then call the transform:

```
val output = slicer.transform(dataset)
output.show(false)
```

Figure 8-21 shows the full VectorSlicer selector.

```scala
scala> val data = Arrays.asList(
     |    Row(Vectors.dense(2.5, 2.9, 3.0)),
     |    Row(Vectors.dense(-2.0, 2.3, 0.0))
     | )
data: java.util.List[org.apache.spark.sql.Row] = [[[2.5,2.9,3.0]], [[-2.0,2.3,0.0]]]

scala>

scala> val defaultAttr = NumericAttribute.defaultAttr
defaultAttr: org.apache.spark.ml.attribute.NumericAttribute = {"type":"numeric"}

scala>

scala> val attrs = Array("col1", "col2", "col3").map(defaultAttr.withName)
attrs: Array[org.apache.spark.ml.attribute.NumericAttribute] =
me":"col3"})

scala>

scala> val attrGroup = new AttributeGroup("InFeatures", attrs.asInstanceOf[Array[Attribute]])
attrGroup: org.apache.spark.ml.attribute.AttributeGroup
3"}]},"num_attrs":3}}

scala>

scala> val dataset = spark.createDataFrame(data, StructType(Array(attrGroup.toStructField())))
dataset: org.apache.spark.sql.DataFrame = [InFeatures: vector]

scala>

scala> val slicer = new VectorSlicer().setInputCol("InFeatures").setOutputCol("SelctedFeatures")
slicer: org.apache.spark.ml.feature.VectorSlicer = vectorSlicer_05c9263c062f

scala>

scala> slicer.setIndices(Array(0))
res54: slicer.type = vectorSlicer_05c9263c062f

scala>

scala> val output = slicer.transform(dataset)
output: org.apache.spark.sql.DataFrame = [InFeatures: vector, SelctedFeatures: vector]

scala> output.show(false)
+--------------+---------------+
|InFeatures    |SelctedFeatures|
+--------------+---------------+
|[2.5,2.9,3.0] |[2.5]          |
|[-2.0,2.3,0.0]|[-2.0]         |
+--------------+---------------+
```

Figure 8-21. *Feature selector VectorSlicer*

Figure 8-22 illustrates the working of the VectorSlicer feature selector.

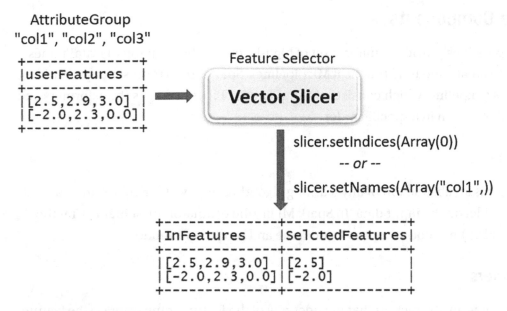

Figure 8-22. *Feature selector VectorSlicer*

ML Pipelines

The `spark.ml` package provides the MLlib APIs for the ML algorithms to create pipelines. The pipeline helps to combine more than one ML algorithm into a single workflow. These are some of the important concepts of ML pipelines.

- *DataFrames*: The ML Sataset can hold variety of data types such as texts, labels, feature vectors in the form of DataFrames through the ML DataFrame APIs. A DataFrame can be created implicitly or explicitly from an RDD. The creation of DataFrames from RDDs was covered in previous chapters.

- *Transformer*: The ML transformer transforms the available DataFrame into another DataFrame. For example, an ML model is a transformer that converts one existing DataFrame into another DataFrame with prediction features.

- *Estimator*: The estimator is an algorithm that helps to create a transformer.

- *Parameters*: The parameters are specified using common APIs for all estimators and transformers.

Pipeline Components

Spark ML pipelines provide a uniform set of high-level APIs built on top of DataFrames that helps to create and tune practical ML pipelines. Spark MLlib represents such a workflow as a pipeline, which consists of a sequence of PipelineStages (transformers and estimators) to be run in a specific order.

Estimators

An estimator is an abstraction of any learning algotithm or any other algorithm that trains the model on the input data. In Spark MLlib, the estimator implements a method `fit()`. The `fit()` method accepts a DataFrame and produces a model.

Transformers

A transformer is an abstraction that includes any of the feature transformers (the feature transformers are explained in the next section of this chapter) and learned models. The transformer implements a method `transform()`, which converts one DataFrame into another, generally by appending one or more columns.

As an example, a feature transformer might take a DataFrame, read a column (e.g., `column1`), map it into a new column (e.g., `column2`), and it gives a new DataFrame as output with the mapped column appended.

Pipeline Examples

The pipeline involves a sequence of algorithms to process and build the model by learning from data. For example, a simple text document processing pipeline might involve the following stages.

1. Split the document's text into the words.

2. Convert each word from the document into a numerical feature vector.

3. Learn from the data and build a prediction model using the feature vectors and the labels.

These steps are the stages in the pipeline. Each stage can be a transformer or an estimator.

Start the Spark Shell (see Figure 8-23) and practice the following code snippets to better understand the pipeline concepts.

```
Spark session available as 'spark'.
Welcome to

    ____              __
   / __/__  ___ _____/ /__
  _\ \/ _ \/ _ `/ __/  '_/
 /___/ .__/\_,_/_/ /_/\_\   version 2.2.1
    /_/

Using Scala version 2.11.8 (Java HotSpot(TM) 64-Bit Server VM, Java 1.8.0_77)
Type in expressions to have them evaluated.
Type :help for more information.

scala>
```

Figure 8-23. *Starting the Spark Shell*

Import the following classes (see Figure 8-24):

```
import org.apache.spark.ml.Pipeline
import org.apache.spark.ml.PipelineModel
import org.apache.spark.ml.classification.LogisticRegression
import org.apache.spark.ml.feature.HashingTF
import org.apache.spark.ml.feature.Tokenizer
import org.apache.spark.ml.linalg.Vector
import org.apache.spark.sql.Row
```

```
scala> import org.apache.spark.ml.{Pipeline, PipelineModel}
import org.apache.spark.ml.{Pipeline, PipelineModel}

scala> import org.apache.spark.ml.classification.LogisticRegression
import org.apache.spark.ml.classification.LogisticRegression

scala> import org.apache.spark.ml.feature.{HashingTF, Tokenizer}
import org.apache.spark.ml.feature.{HashingTF, Tokenizer}

scala> import org.apache.spark.ml.linalg.Vector
import org.apache.spark.ml.linalg.Vector

scala> import org.apache.spark.sql.Row
import org.apache.spark.sql.Row
```

Figure 8-24. *Importing the pipeline APIs from the spark.ml package*

217

> **Note**   The details and workings of logistic regression algorithms are explained later this chapter. We used logistic regression to simply explain the stages of transformers and estimators in a pipeline.

Now prepare the data to train the model with a list of (`id`, `text`, `label`) tuples. The following data set explains the text and the respective label for each text (see Figure 8-25).

```
Schema: ("id", "text", "label")
val training = spark.createDataFrame(Seq(
            (0L, "This is spark book", 1.0),
            (1L, "published by Apress publications", 0.0),
            (2L, "authors are Dharanitharan", 1.0),
            (3L, "and Subhashini", 0.0))).toDF("id", "text", "label")
```

```
scala> val training = spark.createDataFrame(Seq(
     |      (0L, "This is spark book", 1.0),
     |      (1L, "published by Apress publications", 0.0),
     |      (2L, "authors are Dharanitharan", 1.0),
     |      (3L, "and Subhashini", 0.0))).toDF("id", "text", "label")
```

Figure 8-25. *Preparing input documents to train the model*

Now create a pipeline (see Figure 8-26) with three stages: Tokenizer, HashingTF, and the logistic regression algorithm.

```
val tokenizer = new Tokenizer().setInputCol("text").setOutputCol("words")
val hashingTF = new HashingTF().setNumFeatures(1000)
                            .setInputCol(tokenizer.getOutputCol)
                            .setOutputCol("features")
val logitreg = new LogisticRegression().setMaxIter(10).setRegParam(0.001)
val pipeline = new Pipeline().setStages(Array(tokenizer, hashingTF,
logitreg))
```

```
scala> val tokenizer = new Tokenizer().setInputCol("text").
     | setOutputCol("words")
tokenizer: org.apache.spark.ml.feature.Tokenizer = tok_c0f5f43f

scala> val hashingTF = new HashingTF().setNumFeatures(1000).
     | setInputCol(tokenizer.getOutputCol).
     | setOutputCol("features")
hashingTF: org.apache.spark.ml.feature.HashingTF = hashingTF_68

scala> val logitreg = new LogisticRegression().setMaxIter(10).
     | setRegParam(0.001)
logitreg: org.apache.spark.ml.classification.LogisticRegression

scala> val pipeline = new Pipeline().
     | setStages(Array(tokenizer, hashingTF, logitreg))
pipeline: org.apache.spark.ml.Pipeline = pipeline_41712b26b59f
```

Figure 8-26. *Creating the pipeline*

Then, fit the pipeline to the training documents (see Figure 8-27).

```
val model = pipeline.fit(training)
```

```
scala> val model = pipeline.fit(training)
model: org.apache.spark.ml.PipelineModel = pipeline_41712b26b59f
```

Figure 8-27. *Model fitting*

Create the test documents, which are not labeled. We next predict the label based on the feature vectors (see Figure 8-28).

```
val test = spark.createDataFrame(Seq(
          (4L, "spark book"),
          (5L, "apress published this book"),
          (6L, "Dharanitharan wrote this book")))
          .toDF("id", "text")
```

```
scala> val test = spark.createDataFrame(Seq(
     |     (4L, "spark book"),
     |     (5L, "apress published this book"),
     |     (6L, "Dharanitharan wrote this book")
     | )).toDF("id", "text")
test: org.apache.spark.sql.DataFrame = [id: bigint, text: string]
```

Figure 8-28. *Preparing test documents without a label column*

Then make the predictions on the test documents.

```
val transformed = model.transform(test)
```

```
val result = transformed.select("id", "text", "probability", "prediction")
           .collect()
```

```
result.foreach {
        case Row(id: Long, text: String, prob: Vector, prediction:
        Double)
        =>
        println(s"($id, $text) --> prob=$prob, prediction=$prediction")
        }
```

Thus, we have predicted the label based on the feature vectors for each text (see Figure 8-29).

```
scala> val transformed = model.transform(test)
transformed: org.apache.spark.sql.DataFrame

scala> val result = transformed.
      | select("id", "text", "probability", "prediction").
      | collect()

scala> result.
      | foreach{
      | case Row(id: Long, text: String, prob: Vector, prediction: Double)
      | =>
      | println(s"($id, $text) --> prediction=$prediction")
      | }
(4, spark book) --> prediction=1.0
(5, apress published this book) --> prediction=0.0
(6, Dharanitharan wrote this book) --> prediction=1.0
```

Figure 8-29. *Predicting the labels*

Note   The details of Tokenizer and HashingTF transformers were explained earlier in this chapter.

The complete code snippet for the preceding pipeline example is given here.

```scala
package com.apress.pipelines

import org.apache.spark.ml.{Pipeline, PipelineModel}
import org.apache.spark.ml.classification.LogisticRegression
import org.apache.spark.ml.feature.{HashingTF, Tokenizer}
import org.apache.spark.ml.linalg.Vector
import org.apache.spark.sql.Row

object PipelineCreationDemo {

  def main(args: Array[String]): Unit = {

    val sparkSession = SparkSession.builder
                      .appName("PipelineCreationDemo").master("local[*]")
                      .getOrCreate()

    import sparkSession.implicits._

    val training = spark.createDataFrame(Seq(
                (0L, "This is spark book", 1.0),
                (1L, "published by Apress publications", 0.0),
                (2L, "authors are Dharanitharan", 1.0),
                (3L, "and Subhashini", 0.0)))
                .toDF("id", "text", "label")

    val tokenizer = new Tokenizer().setInputCol("text")
                  .setOutputCol("words")

    val hashingTF = new HashingTF().setNumFeatures(1000)
                  .setInputCol(tokenizer.getOutputCol)
                  .setOutputCol("features")

    val logitreg = new LogisticRegression().setMaxIter(10)
                  .setRegParam(0.001)

    val pipeline = new Pipeline()
                  .setStages(Array(tokenizer, hashingTF, logitreg))

    val model = pipeline.fit(training)
```

```scala
    val test = spark.createDataFrame(Seq(
            (4L, "spark book"),
            (5L, "apress published this book"),
            (6L, "Dharanitharan wrote this book")))
             .toDF("id", "text")

    val transformed = model.transform(test)
                .select("id", "text", "probability", "prediction")
                .collect()
result.foreach {
    case Row(id: Long, text: String, prob: Vector, prediction: Double)
      =>
    println(s"($id, $text) --> prob=$prob, prediction=$prediction")
    }
}

}
```

Note   To execute the given code in any IDE that supports Scala, it is mandatory to add the Scala library to the project workspace and all the Spark jars to the classpath.

The working of the discussed simple word text document processing pipeline is illustrated in the flow diagrams that follow.

Figure 8-30 explains the flow of training time usage of pipeline until the fit() method is called. The Pipeline.fit() method is called on the raw data (i.e., original DataFrame), which has raw text documents and labels. The Tokenizer.transform() method splits the raw text documents into words, adding a new column with words to the DataFrame. The HashingTF.transform() method converts the words column into feature vectors, adding a new column with those vectors to the DataFrame. Now, because LogisticRegression is an estimator, the pipeline first calls LogisticRegression.fit() to produce a model; that is, LogisticRegressionModel.

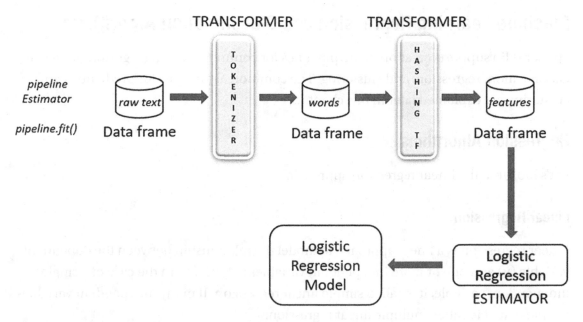

Figure 8-30. *Training time usage of pipeline*

Figure 8-31 explains the flow of PipelineModel, which has the same number of stages as the pipeline. When the PipelineModel's `transform()` method is called on a test data set, the data are passed through the fitted pipeline in order. The `transform()` method in each stage updates the data set and passes it to the next stage. The pipelines and PipelineModels ensure the training and test data go through identical feature processing steps.

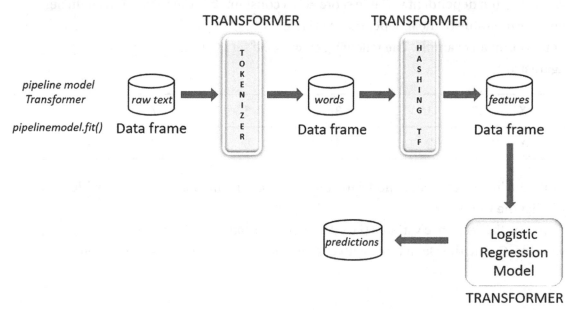

Figure 8-31. *Testing time usage of pipeline model*

Machine Learning Regression and Classification Algorithms

Spark MLlib supports creation of ML pipelines for common learning algorithms such as classification, regression, and clustering. The common algorithms for predictions and clustering are explained later in this chapter.

Regression Algorithms

Let's look into the linear regression approach.

Linear Regression

Linear regression is a linear approach to model the relationship between the dependent variable (y) and one or more independent variables (x1, x2, ...). In the case of a single independent variable, it is called simple linear regression. If many independent variables are present, it is called multiple linear regression.

In linear regression, the linear models are modeled using linear predictor functions whose unknown model parameters are predicted from data. The simple linear regression equation with one dependent and one independent variable is defined by the formula

$$y = a + b(x)$$

where y is the dependent variable score, a is a constant, b is the regression coefficient, and x is the value of the independent variable.

Let's look at an example. The following chart is the set of given observations of y against x.

X	1	3	5	7	9
Y	2	4	6	8	?

Build the linear regression model to build the relationship between the variables to predict the value of x.

Here, y is the response variable (i.e., dependent variable) and x is the independent variable. Create the DataFrame with column labels and features as shown here.

```
val data = List(
            (2.0, Vectors.dense(1.0)),
            (4.0, Vectors.dense(3.0)),
            (6.0, Vectors.dense(5.0)),
            (8.0, Vectors.dense(7.0))
            )
val inputToModel = data.toDF("label","features")
```

where label is the dependent variable (i.e., the value to predict) and the features are the independent variables (i.e., variables used to predict the response variable).

Note   The input DataFrame with label and features to build the model can be created by reading from a file as an RDD and converting it into a DataFrame using the toDF() function.

Now, build the model using LinearRegression().

```
val linearReg = new LinearRegression()
```

```
val linearRegModel = linearReg.fit(inputToModel)
```

The coefficients of the model can be obtained from the coefficients method of the model:

```
println(s"Coefficients:${lrModel.coefficients}
        Intercept:${lrModel.intercept}"
      )
```

Then build the summary of the model.

```
val trainingSummary = linearRegModel.summary
```

```
println(s"numIterations: ${trainingSummary.totalIterations}")
```

```
println(
        s"objectiveHistory:[${trainingSummary.objectiveHistory.
        mkString(",")}]"
        )
```

```
trainingSummary.residuals.show()
```

```
println(s"RMSE: ${trainingSummary.rootMeanSquaredError}")
```

```
println(s"r2: ${trainingSummary.r2}")
```

Now, the label for feature 9.0 can be predicted as:

```
val toPredict = List((0.0,Vectors.dense(9.0)),(0.0,Vectors.dense(11.0)))
```

```
val toPredictDF = toPredict.toDF("label","features")
```

```
val predictions=linearRegModel.transform(toPredictDF)
```

```
predictions.select("prediction").show()
```

Figure 8-32 shows the execution result of each step in building the regression model.

```scala
scala> import org.apache.spark.ml.regression.LinearRegression
import org.apache.spark.ml.regression.LinearRegression

scala> val data = List(
     |    (2.0, Vectors.dense(1.0)),
     |    (4.0, Vectors.dense(3.0)),
     |    (6.0, Vectors.dense(5.0)),
     |    (8.0, Vectors.dense(7.0))
     | )
data: List[(Double, org.apache.spark.ml.linalg.Vector)]

scala> val inputToModel = data.toDF("label","features")
inputToModel: org.apache.spark.sql.DataFrame = [label: double, features: vector]

scala>

scala> val linearReg = new LinearRegression()
linearReg: org.apache.spark.ml.regression.LinearRegression = linReg_31126ef4c5eb

scala>

scala> val linearRegModel = linearReg.fit(inputToModel)
linearRegModel: org.apache.spark.ml.regression.LinearRegressionModel = linReg_3112

scala> println(s"Coefficients:${lrModel.coefficients} Intercept:${lrModel.intercept}")
Coefficients:[1.7922744891641464] Intercept:0.6231765325075609

scala>

scala> val trainingSummary = linearRegModel.summary
trainingSummary: org.apache.spark.ml.regression.LinearRegressionTrainingSummary

scala>

scala> println(s"numIterations: ${trainingSummary.totalIterations}")
numIterations: 1

scala> println(s"objectiveHistory:[${trainingSummary.objectiveHistory.mkString(",")}]")
objectiveHistory:[0.0]

scala>

scala> trainingSummary.residuals.show()
+--------------------+
|           residuals|
+--------------------+
|  -2.66453525910037...|
|  -1.77635683940025...|
|  -1.77635683940025...|
|                 0.0|
+--------------------+


scala>

scala> println(s"RMSE: ${trainingSummary.rootMeanSquaredError}")
RMSE: 1.831026719408895E-15

scala>

scala> println(s"r2: ${trainingSummary.r2}")
r2: 1.0

scala> val toPredict = List(
     |    (0.0,Vectors.dense(9.0)),
     |    (0.0,Vectors.dense(11.0))
     | )
toPredict: List[(Double, org.apache.spark.ml.linalg.Vector)]

scala>

scala> val toPredictDF = toPredict.toDF("label","features")
toPredictDF: org.apache.spark.sql.DataFrame = [label: double, features: vector]

scala>

scala> val predictions=linearRegModel.transform(toPredictDF)
predictions: org.apache.spark.sql.DataFrame

scala>

scala> predictions.select("prediction").show()
+------------------+
|        prediction|
+------------------+
|              10.0|
|11.999999999999998|
+------------------+
```

Figure 8-32. *Linear regression algorithm*

Thus, the values of 9 and 11 have been predicted as 10.0 and 11.998 (~=12 approx), respectively.

The complete code snippet for the regression algorithm implementation is given here.

```
package com.apress.mlalgorithms

import org.apache.spark.ml_
import org.apache.spark.ml.linalg.Vector
import org.apache.spark.ml.linalg.Vectors
import org.apache.spark.ml.regression.LinearRegression
import org.apache.spark.sql.SparkSession

object LinearRegressionDemo {

  def main(args: Array[String]): Unit = {

    val sparkSession = SparkSession.builder
                     .appName("LinearRegressionDemo").master("local[*]")
                     .getOrCreate()

    import sparkSession.implicits._

    val data = List(
             (2.0, Vectors.dense(1.0)),
             (4.0, Vectors.dense(3.0)),
             (6.0, Vectors.dense(5.0)),
             (8.0, Vectors.dense(7.0))
             )
    val inputToModel = data.toDF("label","features")
    val linearReg = new LinearRegression()
    val linearRegModel = linearReg.fit(inputToModel)
    println(s"Coefficients:${lrModel.coefficients}
          Intercept:${lrModel.intercept}")
    val trainingSummary = linearRegModel.summary
    println(s"numIterations: ${trainingSummary.totalIterations}")
    println(s"objectiveHistory:[
          ${trainingSummary.objectiveHistory.mkString(",")}]")
    trainingSummary.residuals.show()
```

```
  println(s"RMSE: ${trainingSummary.rootMeanSquaredError}")
  println(s"r2: ${trainingSummary.r2}")
  val toPredict = List((0.0,Vectors.dense(9.0)),
                       (0.0,Vectors.dense(11.0)))
  val toPredictDF = toPredict.toDF("label","features")
  val predictions=linearRegModel.transform(toPredictDF)
  predictions.select("prediction").show()
  }
}
```

Classification Algorithms

Let's now look into the logistic regression approach.

Logistic Regression

The logistic regression is used to predict the categorical response. The `spark.ml` logistic regression can be used to predict a binary outcome (either 0 or 1) by using binomial logistic regression.

The following example shows how to train binomial logistic regression models for binary classification to predict the categorical response. Create the data set shown in Figure 8-33 in a file `matchPlay.csv`.

```
outlook,temp,humidity,played
sunny,hot,high,0
sunny,hot,high,0
overcast,hot,high,1
rainy,mild,high,1
rainy,cool,normal,1
rainy,cool,normal,0
overcast,cool,normal,1
sunny,mild,high,0
sunny,cool,normal,1
rainy,mild,normal,1
sunny,mild,normal,1
overcast,mild,high,1
overcast,hot,normal,1
rainy,mild,high,0
```

Figure 8-33. matchPlay.csv file

The data set contains four variables: outlook, temp, humidity, and play. They explain whether the match is played or not based on outlook, temperature, and humidity conditions. Play is the response variable and the other three columns are independent variables.

Now, build a logistic regression model to predict whether the match would be played or not based on the independent variables' labels.

First, read the data from the file using Spark Session.

```
val data = spark.read.option("header","true")
                      .option("inferSchema","true")
                      .format("csv")
                      .load("matchPlay.txt")
```

Verify the schema using data.printSchema(). Select the required label columns and the feature columns. Here the label column is "played", as it is the response variable, and the other columns are feature columns, which helps for the prediction.

```
val logRegDataAll = (data.select(data("play").as("label"),
                     $"outlook",$"temp",$"humidity"))
```

Next, convert the categorical (i.e., string) columns into numerical values because the ML algorithm cannot understand the categorical variable. This can be done using StringIndexer, which creates the column of indices from the column of labels. The StringIndexer was explained earlier in this chapter.

```
import org.apache.spark.ml.feature.StringIndexer

val outlookIndexer = new StringIndexer()
                       .setInputCol("outlook").setOutputCol("OutlookIndex")

val tempIndexer = new StringIndexer()
                       .setInputCol("temp").setOutputCol("tempIndex")

val humidityIndexer = new StringIndexer()
                       .setInputCol("humidity").
                       setOutputCol("humidityIndex")
```

Third, apply OneHotEncoder (i.e., 0 or 1) to the numerical values. One-hot encoding maps a categorical feature, represented as a label index, to a binary vector with at most a single one-value by indicating the presence of a specific feature value from the set of all feature values.

Because the categorical feature is represented as a label index, we need to map the label index to a binary vector with at most a single one-value indicating the presence of a specific feature value from among the set of all feature values. OneHotEncoder is also a transformer, which can be used in the ML pipeline.

```
import org.apache.spark.ml.feature.OneHotEncoder

val outlookEncoder = new OneHotEncoder()
                    .setInputCol("OutlookIndex").
                    setOutputCol("outlookVec")

val tempEncoder = new OneHotEncoder()
                    .setInputCol("tempIndex").setOutputCol("tempVec")

val humidityEncoder = new OneHotEncoder()
                    .setInputCol("humidityIndex").
                    setOutputCol("humidityVec")
```

Fourth, create the label and features to build the model by assembling the OneHotEncoded vectors of all the categorical columns to the features vector.

```
import org.apache.spark.ml.linalg.Vectors
import org.apache.spark.ml.feature.VectorAssembler

val assembler = new VectorAssembler()
                    .setInputCols(Array("outlookVec","tempVec","humidity
                    Vec"))
                    .setOutputCol("features")
```

Fifth, create a LogisticRegression estimator to build the pipeline.

```
import org.apache.spark.ml.Pipeline
import org.apache.spark.ml.classification.LogisticRegression

val logReg = new LogisticRegression()

val pipeline = new Pipeline().setStages
                    (
                    Array(outlookIndexer,tempIndexer,humidityIndexer,
                    outlookEncoder,tempEncoder,humidityEncoder,
                    assembler,logReg)
                    )
```

Sixth, randomly split the original data set into training (70%) and test (30%) to build the logistic regression model and verify it with the predicted label for the "played" variable.

```
val model = pipeline.fit(training)

val results = model.transform(test)

results.select("outlook","humidity","temp","label","prediction").show()
```

Note   Execute the statements and observe the flow of the pipeline.

The complete code for the example just described is given here and the model is shown in Figure 8-34.

```
package com.apress.mlalgorithms

import org.apache.spark.ml.classification.LogisticRegression
import org.apache.spark.ml.Pipeline
import org.apache.spark.ml.feature.{VectorAssembler, StringIndexer}
import org.apache.spark.ml.feature.{VectorIndexer,OneHotEncoder}
import org.apache.spark.ml.linalg.Vectors

object LogisticRegressionDemo {

  def main(args: Array[String]): Unit = {

    val sparkSession = SparkSession.builder
                     .appName("LogisticRegression").master("local[*]")
                     .getOrCreate()

    import sparkSession.implicits._

    val data = spark.read.option("header","true")
                  .option("inferSchema","true").format("csv")
                  .load("matchPlay.txt")

    val logRegDataAll = (data.select(data("play")
                      .as("label"),$"outlook",$"temp",$"humidity"))

    // converting string column to numerical values
```

```scala
val outlookIndexer = new StringIndexer().setInputCol("outlook")
                    .setOutputCol("OutlookIndex")

val tempIndexer = new StringIndexer().setInputCol("temp")
                    .setOutputCol("tempIndex")

val humidityIndexer = new StringIndexer().setInputCol("humidity")
                    .setOutputCol("humidityIndex")

// converting numerical values into OneHot Encoding - 0 or 1

val outlookEncoder = new OneHotEncoder().setInputCol("OutlookIndex")
                    .setOutputCol("outlookVec")

val tempEncoder = new OneHotEncoder().setInputCol("tempIndex")
                    .setOutputCol("tempVec")

val humidityEncoder = new OneHotEncoder().setInputCol("humidityIndex")
                    .setOutputCol("humidityVec")

// create(label, features)

val assembler = new  VectorAssembler()
            .setInputCols(Array("outlookVec","tempVec","humidityVec"))
            .setOutputCol("features")

val Array(training,test)=logRegDataAll.randomSplit(Array(0.7,0.3))

val logReg = new LogisticRegression()

val pipeline = new Pipeline()
                .setStages(Array(outlookIndexer,tempIndexer,
                humidityIndexer,outlookEncoder,tempEncoder,
                humidityEncoder,assembler,logReg))

val model = pipeline.fit(training)

val results = model.transform(test)

results.select("outlook","humidity","temp","label","prediction").show()

  }
}
```

```
scala> val model = pipeline.fit(training)
model: org.apache.spark.ml.PipelineModel = pipeline_08375a00f852

scala>

scala> val results = model.transform(test)
results: org.apache.spark.sql.DataFrame

scala>

scala> results.select("outlook","humidity","temp","label","prediction").show()
+--------+--------+----+-----+----------+
| outlook|humidity|temp|label|prediction|
+--------+--------+----+-----+----------+
|   rainy|  normal|cool|    0|       1.0|
|   sunny|    high| hot|    0|       0.0|
|overcast|    high| hot|    1|       0.0|
|   rainy|  normal|cool|    1|       1.0|
+--------+--------+----+-----+----------+
```

Figure 8-34. *Logistic regression model*

Clustering Algorithms

Let's look into the K-Means clustering algorithm.

K-Means Clustering

The K-Means clustering algorithm is used to cluster the data points into a preferred number of clusters. In Spark MLlib, K-Means is implemented as an estimator and generates a K MeansModel as a base model.

The details of the input columns and output columns are described next.

- Input columns

  - *Parameter name:* featuresCol

  - *Type(s)*: Vector

  - *Default:* "features", which is a feature vector

- Output columns

  - *Parameter name*: predictionCol

  - *Type(s)*: Int

  - *Default*: "prediction", which is the predicted cluster center

As an example, create the data set shown in Figure 8-35 in a file called kmeans-sample.txt.

```
0 1:0.0 2:0.0 3:0.0
1 1:0.1 2:0.1 3:0.1
2 1:0.2 2:0.2 3:0.2
3 1:9.0 2:9.0 3:9.0
4 1:9.1 2:9.1 3:9.1
5 1:9.2 2:9.2 3:9.2
```

Figure 8-35. *kmeans-sample.txt file*

Import the classes for K-Means clustering. The model is shown in Figure 8-36.

```
import org.apache.spark.ml.clustering.KMeans

// Load the dataset in "libsvm" format
val dataset = spark.read.format("libsvm").load("kmeans-sample.txt ")

// Trains a k-means model by setting the number of clusters as 2.
val kmeans = new KMeans().setK(2).setSeed(1L)
val model = kmeans.fit(dataset)

// Make predictions
val predictions = model.transform(dataset)

// print the result.
model.clusterCenters.foreach(println)
```

```
scala> import org.apache.spark.ml.clustering.KMeans
import org.apache.spark.ml.clustering.KMeans

scala> val dataset = spark.read.format("libsvm").load("kmeans-sample.txt")
dataset: org.apache.spark.sql.DataFrame = [label: double, features: vector]

scala> val kmeans = new KMeans().setK(2).setSeed(1L)
kmeans: org.apache.spark.ml.clustering.KMeans = kmeans_05f9164319cf

scala> val model = kmeans.fit(dataset)
model: org.apache.spark.ml.clustering.KMeansModel = kmeans_05f9164319cf

scala> val predictions = model.transform(dataset)
predictions: org.apache.spark.sql.DataFrame = [label: double, features: vector

scala> model.clusterCenters.foreach(println)
[0.1,0.1,0.1]
[9.1,9.1,9.1]
```

Figure 8-36. *K-Means clustering model*

Points to Remember

- Spark MLlib is Spark's collection of ML libraries, which can be used as APIs to implement ML algorithms.

- Use the common learning algorithms such as classification, regression, clustering, and collaborative filtering.

- Construct, evaluate, and tune the ML pipelines using Spark MLlib.

- In ML pipelines, extraction deals with extracting the features with the raw data.

- Transformation deals with scaling, converting, and modifying the features extracted from the raw data.

- Selection deals with taking a sample or subset from a larger set of features.

In the next chapter, we discuss the features of SparkR.

CHAPTER 9

Working with SparkR

In the previous chapter, we discussed the fundamental concepts of Spark MLlib. We also discussed the machine learning algorithms with implementation.

In this chapter, we are going to discuss how to work with the SparkR component. We focus on the following topics:

- Introduction to SparkR.

- Starting SparkR from RStudio.

- Creating a SparkDataFrame.

- SparkDataFrame operations.

- Applying user-defined functions.

- Running SQL queries.

Introduction to SparkR

SparkR is an R package that allows us to use Apache Spark from R. Spark provides a distributed DataFrame that is like R data frames to perform select, filter, and aggregate operations on large data sets. SparkR also supports distributed ML algorithms using MLlib.

SparkDataFrame

A SparkDataFrame is a distributed collection of data organized into named columns. A SparkDataFrame is equivalent to a table in an RDBMS or a data frame in R with richer optimization under the hood. SparkDataFrame can be constructed from different sources, such as structure data files, external databases, tables in Hive, existing local R data frames.

237

© Subhashini Chellappan, Dharanitharan Ganesan 2018
S. Chellappan and D. Ganesan, *Practical Apache Spark*, https://doi.org/10.1007/978-1-4842-3652-9_9

SparkSession

The entry point for SparkR is the SparkSession. The SparkSession connects the R program to a Spark cluster. The `spark.session` is used to create SparkSession. You can also pass options such as application name, dependent Spark packages, and so on, to the `spark.session`.

Note If you are working from the SparkR shell, the SparkSession should already be created for you, and you would not need to call `sparkR.session`.

Let's discuss how to start SparkR from RStudio.

Starting SparkR from RStudio

1. Download Spark version 2.3.0 from this link.

   `http://www-us.apache.org/dist/spark/spark-2.3.0/spark-2.3.0-bin-hadoop2.7.tgz`

2. Extract the tar to `spark_2.3.0`.

3. Download R from this link and install it.

   `https://cran.r-project.org/bin/windows/base/old/3.4.2/`

4. Download RStudio from this link and install it.

   `https://www.rstudio.com/products/rstudio/download/`

5. Start RStudio.

6. Install SparkR packages by issuing the following command (see Figure 9-1).

   `Library.packages(SparkR)`

```
R version 3.4.4 (2018-03-15) -- "Someone to Lean On"
Copyright (C) 2018 The R Foundation for Statistical Computing
Platform: x86_64-w64-mingw32/x64 (64-bit)

R is free software and comes with ABSOLUTELY NO WARRANTY.
You are welcome to redistribute it under certain conditions.
Type 'license()' or 'licence()' for distribution details.

R is a collaborative project with many contributors.
Type 'contributors()' for more information and
'citation()' on how to cite R or R packages in publications.

Type 'demo()' for some demos, 'help()' for on-line help, or
'help.start()' for an HTML browser interface to help.
Type 'q()' to quit R.

[Workspace loaded from ~/.RData]

> install.packages("SparkR")
Installing package into 'C:/Users/r.c.subhashini/Documents/R/win-library/3.4'
(as 'lib' is unspecified)
trying URL 'https://cran.rstudio.com/bin/windows/contrib/3.4/SparkR_2.3.0.zip'
Content type 'application/zip' length 1595574 bytes (1.5 MB)
downloaded 1.5 MB

package 'SparkR' successfully unpacked and MD5 sums checked

The downloaded binary packages are in
        C:\Users\r.c.subhashini\AppData\Local\Temp\Rtmpw8hCch\downloaded_packages
> |
```

Figure 9-1. *Installing SparkR packages*

7. Attach the SparkR package to the R environment by calling this
    command (see Figure 9-2).

    ```
    library(SparkR)
    ```

```
> library(SparkR)

Attaching package: 'SparkR'

The following objects are masked from 'package:stats':

    cov, filter, lag, na.omit, predict, sd, var, window

The following objects are masked from 'package:base':

    as.data.frame, colnames, colnames<-, drop, endsWith, intersect, rank, rbind, sample, startsWith, subset,
    summary, transform, union

> |
```

Figure 9-2. *Attaching the SparkR package*

8. Set the Spark environment variable by issuing these commands
   (see Figure 9-3).

```
if (nchar(Sys.getenv("SPARK_HOME")) < 1) {
  Sys.setenv(SPARK_HOME = "C:/Users/Administrator/Desktop/
  spark-2.3.0")
}
```

```
> if (nchar(Sys.getenv("SPARK_HOME")) < 1) {
+     Sys.setenv(SPARK_HOME = "C://Users//Administrator//Desktop//spark_2.3.0")}
> |
```

Figure 9-3. *Setting the Spark environment variable*

9. Load SparkR and call `sparkR.session` by issuing these
   commands. You can also specify Spark driver properties (see
   Figure 9-4).

```
library(SparkR, lib.loc = c(file.path(Sys.getenv("SPARK_HOME"),
"R", "lib")))
sparkR.session(master = "local[*]",sparkHome =Sys.getenv("SPARK_
HOME"),enableHiveSupport = TRUE,  sparkConfig = list(spark.driver.
memory = "2g"))
```

```
> library(SparkR, lib.loc = c(file.path(Sys.getenv("SPARK_HOME"), "R", "lib")))
> sparkR.session(master = "local[*]",sparkHome =Sys.getenv("SPARK_HOME"),enableHiveSupport = TRUE,  sparkConfig = list(spark.driver.memory
 = "2g"))
Spark package found in SPARK_HOME: C://Users//Administrator//Desktop//spark-2.3.0
Launching java with spark-submit command C://Users//Administrator//Desktop//spark-2.3.0/bin/spark-submit2.cmd   --driver-memory "2g" spark
r-shell C:\Users\RC8A06~1.SUB\AppData\Local\Temp\RtmpS24tZ4\backend_port1ef876e94462
Java ref type org.apache.spark.sql.SparkSession id 1
> |
```

Figure 9-4. *Load SparkR and call* `sparkR.session`

We have successfully created a SparkR session.

Creating SparkDataFrames

There are three ways to create SparkDataFrames:

- From a local R DataFrame.
- From a Hive table.
- From other data sources.

Let's discuss each method in turn.

From a Local R DataFrame

The easiest way to create SparkDataFrame is to convert a local R DataFrame into a SparkDataFrame. You can use `as.DataFrame` or `createDataFrame` to create a SparkDataFrame. The following code creates a SparkDataFrame using a faithful data set from R (see Figure 9-5).

```
df <- as.DataFrame(faithful)                        //Line1

# To display the first part of the SparkDataFrame
head(df)                                            //Line 2
```

```
> df <- as.DataFrame(faithful)
> head(df)
  eruptions waiting
1     3.600      79
2     1.800      54
3     3.333      74
4     2.283      62
5     4.533      85
6     2.883      55
> |
```

Figure 9-5. Creating a SparkDataFrame from a local R DataFrame

From Other Data Sources

The read.df method is the general method to create a SparkDataFrame from data sources. SparkR also supports JSON, CSV, and Parquet files. The following code creates a SparkDataFrame from a JSON file (see Figure 9-6 for authors.json).

```
{"name":"Subhashini", "publiation":"Apress"}
{"name":"Dharanidhran", "publiation":"Apress"}
```

Figure 9-6. *authors.json*

```
authors <- read.df ("C://SparkR//authors.json", "json")    → Line 1
head(authors)                                              → Line 2
```

The output is shown in Figure 9-7.

```
> authors <- read.df ("C://SparkR//authors.json", "json")
> head(authors)
          name publiation
1    Subhashini      Apress
2 Dharanidhran      Apress
```

Figure 9-7. *Creating a SparkDataFrame using a JSON file*

Let's discuss how to create a SparkDataFrame from a .csv file. Refer to Figure 9-8 for the authors.csv file.

```
name,authors
Subhashini,Apress
Dharanidhran,Apress
```

Figure 9-8. *authors.csv file*

The following code creates a SparkDataFrame from a `.csv` file. The output is shown in Figure 9-9.

```
csvdf <- read.df("C://SparkR//authors.csv", "csv", header = "true",
inferSchema = "true", na.strings = "NA")                    → Line 1
head(csvdf)                                                  → Line 2
```

```
> csvdf <- read.df("C://SparkR//authors.csv", "csv", header = "true", inferSchema = "true", na.strings = "
NA")
>
> head(csvdf)
          name authors
1    Subhashini  Apress
2 Dharanidhran  Apress
> |
```

Figure 9-9. *Creating a SparkDataFrame using a `.csv` file*

From Hive Tables

To create a SparkDataFrame from a Hive table we need to create a SparkSession with Hive support (`enableHiveSupport = TRUE`) to access tables in the Hive metastore. The following code creates a SparkDataFrame from a Hive table. The output is shown in Figure 9-10.

Create hive table authors and load data into authors table

```
sql("CREATE TABLE IF NOT EXISTS authors (name STRING, publication STRING)
ROW FORMAT DELIMITED FIELDS TERMINATED BY ','")
sql("LOAD DATA LOCAL INPATH 'C:/SparkR/authors.csv' INTO TABLE authors")

# Queries can be expressed in HiveQL.
results <- sql("FROM authors SELECT name,publication")

# results is now a SparkDataFrame
head(results)
```

```
> sql("CREATE TABLE IF NOT EXISTS authors (name STRING, publication STRING) ROW FORMAT DELIMITED FIELDS TERMINATED BY ','")
SparkDataFrame[]
> sql("LOAD DATA LOCAL INPATH 'C:/SparkR/authors.csv' INTO TABLE authors")
SparkDataFrame[]
> results <- sql("FROM authors SELECT name,publication")
> head(results)
          name publication
1    Subhashini      Apress
2 Dharanidhran      Apress
> |
```

Figure 9-10. *Creating a SparkFrame from a Hive table*

SparkDataFrame Operations

SparkDataFrames support several functions to perform structured data processing.

Selecting Rows and Columns

Let us consider the SparkDataFrame result created from the Hive table `authors`.

```
# To get the basic information about the SparkDataFrame
```

```
results
```

Figure 9-11 shows the `results` SparkDataFrame.

```
# To select only the "name" column
```

```
head(select(results, results$name))
```

```
> results
SparkDataFrame[name:string, publication:string]
> |
```

Figure 9-11. `results` *SparkDataFrame*

Figure 9-12 shows the `select` query output.

```
# You can also pass in column name as strings
```

```
head(select(results, "name"))
```

```
> head(select(results, results$name))
          name
1     Subhashini
2 Dharanidhran
> |
```

Figure 9-12. *Output of* `select` *query*

To apply filter condition to the SparkDataFrame (see Figure 9-13).

```
head(filter(results, results$name == 'Subhashini'))
```

```
> head(filter(results, results$name == 'Subhashini'))
          name publication
1 Subhashini       Apress
> head(filter(results, results$name == 'Dharanidharan'))
[1] name           publication
<0 rows> (or 0-length row.names)
> |
```

Figure 9-13. *Output of* select *query with condition*

Grouping and Aggregation

SparkR DataFrames supports many functions to aggregate data after grouping. Let's consider the student.csv file shown in Figure 9-14.

1001	John	45.0
1002	James	85.0
1003	John	45.0
1004	James	85.0
1005	Smith	60.0
1006	Scott	70.0
1007	Shoba	80.0
1008	Taanu	90.0
1009	Anbu	95.0
1010	Aruna	85.0

Figure 9-14. student.csv *file*

The following code creates a student DataFrame from a `.csv` file.

```
students <- read.df("C://SparkR//student.csv", "csv", header = "true",
inferSchema = "true", na.strings = "NA")

# Use n operator to count the number of times each grade appears

head(summarize(groupBy(students, students$grade), count =
n(students$grade)))
```

The output grade count is shown in Figure 9-15.

```
> students <- read.df("C://SparkR//student.csv", "csv", header = "true", inferSchema = "true", na.strings = "NA")
> head(summarize(groupBy(students, students$grade), count = n(students$grade)))
  grade count
1    70     1
2    80     1
3    85     3
4    45     2
5    60     1
6    95     1
> |
```

Figure 9-15. *Grade count output*

```
# Sort the output from the aggregation to get the most common grade.

grade_counts <- summarize(groupBy(students, students$grade), count =
n(students$grade))

head(arrange(grade_counts, desc(grade_counts$count)))
```

The common grade output is shown in Figure 9-16.

```
> grade_counts <- summarize(groupBy(students, students$grade), count = n(students$grade))
>
> head(arrange(grade_counts, desc(grade_counts$count)))
  grade count
1    85     3
2    45     2
3    60     1
4    80     1
5    95     1
6    90     1
>
> |
```

Figure 9-16. *The common grade output*

Let's see how to find the average grade.

```
head(select(students, avg(students$grade)))
```

The output is shown in Figure 9-17.

```
> head(select(students, avg(students$grade)))
  avg(grade)
1         74
>
```

Figure 9-17. *Average grade*

Operating on Columns

SparkR also provides functions that can be directly applied to columns for data processing.

```
# Add 5 marks to the grade column.
```

```
# To assign this to a new column in the same SparkDataFrame
```

```
students$new_grade <- students$grade + 5
```

```
head(students)
```

The output is shown in Figure 9-18.

```
> students$new_grade <- students$grade + 5
>
> head(students)
  studId   name grade new_grade
1   1001   John    45        50
2   1002  James    85        90
3   1003   John    45        50
4   1004  James    85        90
5   1005  Smith    60        65
6   1006  Scott    70        75
>
```

Figure 9-18. *Adding new_grade function*

Applying User-Defined Functions

SparkR supports several kinds of user-defined functions.

Run a Given Function on a Large Data Set Using dapply or dapplyCollect

Use the dapply function to apply a function to each partition of a SparkDataFrame. The function takes one parameter, a data.frame that corresponds to each partition. The output of the function should be a data.frame. The Schema specifies the row format of the resulting SparkDataFrame. It should match the data types of returned values. The following code adds five marks to the grade column.

students_details <- read.df("C://SparkR//student.csv", "csv", header = "true", inferSchema = "true", na.strings = "NA")

```
# create a schema.
schema <- structType(structField("studId", "int"), structField("studName",
"string"),
                     structField("grade", "double"), structField("new_
                     grade", "double"))

students_new_grade <- dapply(students_details, function(x) { x <- cbind(x,
x$grade + 5) }, schema)


head(collect(students_new_grade))
```

The output is shown in Figure 9-19.

```
> students_details <- read.df("C://SparkR//student.csv", "csv", header = "true", inferSchema = "true", na.strings = "NA")
> schema <- structType(structField("studId", "int"), structField("studName", "string"),
+                      structField("grade", "double"), structField("new_grade", "double"))
>
> students_new_grade <- dapply(students_details, function(x) { x <- cbind(x, x$grade + 5) }, schema)
> head(collect(students_new_grade))
  studId studName grade new_grade
1   1001     John    45        50
2   1002    James    85        90
3   1003     John    45        50
4   1004    James    85        90
5   1005    Smith    60        65
6   1006    Scott    70        75
> |
```

Figure 9-19. dapply

The dapplyCollect function is like dapply. It applies a function to each partition of a SparkDataFrame and collects the result back. The output of this function should be a DataFrame. However, schema is not required to be passed.

```
students_new_grade <- dapplyCollect(
                    students_details,
                    function(x) {
                    x <- cbind(x, "new_grade" = x$grade + 5)
         })
head(students_new_grade, 3)
```

Note   dapplyCollect can fail if the output of UDF (i.e., User Defined Function) run on all the partitions cannot be pulled to the driver and fit in driver memory.

The output is shown in Figure 9-20.

```
> students_new_grade <- dapplyCollect(
+       students_details,
+       function(x) {
+           x <- cbind(x, "new_grade" = x$grade + 5)
+       })
> head(students_new_grade, 3)
  studId   name grade new_grade
1   1001   John    45        50
2   1002  James    85        90
3   1003   John    45        50
>
> |
```

Figure 9-20. dapplyCollect

Running SQL Queries from SparkR

Let's discuss how to run SQL queries from SparkR. We can register a SparkDataFrame as a temporary view in Spark SQL and run SQL queries over its data. It returns the result as a SparkDataFrame.

```
# Load a JSON file

authors <- read.df("C://SparkR/authors.json", "json")

# Register this SparkDataFrame as a temporary view.

createOrReplaceTempView(authors, "authors")
```

```
# SQL statements can be run by using the sql method

result <- sql("SELECT name FROM authors")

head(result)
```

The output is shown in Figure 9-21.

```
> authors <- read.df("C://SparkR/authors.json", "json")
> createOrReplaceTempView(authors, "authors")
> result <- sql("SELECT name FROM authors")
> head(result)
        name
1    Subhashini
2 Dharanidhran
> |
```

Figure 9-21. SQL query output

Machine Learning Algorithms

Spark R supports various supervised and unsupervised machine algorithms. We have already learned linear regression, logistic regression, and clustering algorithms in the previous chapter and implemented the same as Spark ML pipelines. In this chapter, we discuss the implementation of the same algorithms using SparkR libraries.

Regression and Classification Algorithms

Let's discuss regression and classification algorithms.

Linear Regression

The spark.glm {SparkR} package is used to fit the generalized linear model against the SparkDataFrame.

Usage: spark.glm(data, formula, family)

data: The SparkDataFrame for training the model.

formula: A symbolic description of the model to be fitted. The operators '~', '.', ':', '+', and '-' are supported by the model.

family: The description of the error distribution and link function to be used in the model.

The simple linear regression equation with one dependent and one independent variable is defined by the formula

$$y = a + b(x)$$

where y is the dependent variable score, a is a constant, b is the regression coefficient, and x is the value of an independent variable.

`spark.glm` returns a fitted generalized linear model.

```
spark.glm(dataFrame, y~x)
```

The `summary` and `predict` methods are available for the fitted model and their usage is described next.

```
summary(GeneralizedLinearRegressionModel)
predict(GeneralizedLinearRegressionModel)
```

Also, the method `write.ml(model,path)` can be used to save the fitted model to any path that can be loaded again and used later.

Let's look at an example. Table 9-1 presents the set of given observations of y against x.

Table 9-1. *Observation of y against x*

x	1	3	5	7	9
y	2	4	6	8	?

Build the linear regression model to build the relationship between the variables to predict the value of x. Here, y is the response variable (i.e., dependent variable) and x is the independent variable.

Start the R environment and create the `spark.session` as discussed earlier in this chapter.

```
if (nchar(Sys.getenv("SPARK_HOME")) < 1) {
  Sys.setenv(SPARK_HOME = "C:/Users/Administrator/Desktop/spark-2.3.0")
}
```

Load the Spark libraries as shown here (see Figure 9-22).

```
library(SparkR, lib.loc = c(file.path(Sys.getenv("SPARK_HOME"), "R", "lib")))

sparkR.session(master = "local[*]",
               sparkHome =Sys.getenv("SPARK_HOME"),
               enableHiveSupport = FALSE,
               sparkConfig = list(spark.driver.memory = "2g")
               )

x <- c(1,3,5,7)

y <- c(2,4,6,8)

dataFrameInR <- data.frame(x=c(1,3,5,7),y=c(2,4,6,8))

sparkDataFrame <- createDataFrame(dataFrameInR)
```

```
if (nchar(Sys.getenv("SPARK_HOME")) < 1) {
  Sys.setenv(SPARK_HOME = "C:/Users/Administrator/Desktop/spark-2.3.0")
}

library(SparkR, lib.loc = c(file.path(Sys.getenv("SPARK_HOME"), "R", "lib")))

sparkR.session(master = "local[*]",
               sparkHome =Sys.getenv("SPARK_HOME"),
               enableHiveSupport = FALSE,
               sparkConfig = list(spark.driver.memory = "2g")
               )

x <- c(1,3,5,7)

y <- c(2,4,6,8)

dataFrameInR <- data.frame(x=c(1,3,5,7),y=c(2,4,6,8))

sparkDataFrame <- createDataFrame(dataFrameInR)
```

```
> print(sparkDataFrame)
SparkDataFrame[x:double, y:double]
```

Figure 9-22. *Spark DataFrame from linear model*

Now create the linear model using the `glm` package as shown here.

```
linearModel <- spark.glm(sparkDataFrame, y ~ x, family = "gaussian")
```

Use the `summary` function to print the summary of the created model, as shown in Figure 9-23.

```
> summary(linearModel)

Deviance Residuals:
(Note: These are approximate quantiles with relative error <= 0.01)
        Min            1Q         Median             3Q            Max
-2.6645e-15   -2.6645e-15   -1.7764e-15   -1.7764e-15    0.0000e+00

Coefficients:
              Estimate   Std. Error     t value   Pr(>|t|)
(Intercept)          1   2.6534e-15   3.7687e+14          0
x                    1   5.7902e-16   1.7271e+15          0

(Dispersion parameter for gaussian family taken to be 6.705318e-30)

    Null deviance: 2.0000e+01   on 3   degrees of freedom
Residual deviance: 1.3411e-29   on 2   degrees of freedom
AIC: -254.1

Number of Fisher Scoring iterations: 1
```

Figure 9-23. *Summary of linear model*

This summary of the linear model shows that the coefficient values are a=1 and b=1. Hence the linear relationship between y and x is y ~ 1 + x. Now use the `predict` function to predict the y values for any x values as shown in Figure 9-24.

```
dataFrametoPredict <- data.frame(x=c(9,11,13))

sparkDataFrameToPredict <- createDataFrame(dataFrametoPredict)

fittedModel <- predict(linearModel, sparkDataFrameToPredict)

head(select(fittedModel, "prediction"))
```

```
dataFrametoPredict <- data.frame(x=c(9,11,13))

sparkDataFrameToPredict <- createDataFrame(dataFrametoPredict)

fittedModel <- predict(linearModel, sparkDataFrameToPredict)

head(select(fittedModel, "prediction"))
```

```
    > head(select(fittedModel, "prediction"))
      prediction
    1         10
    2         12
    3         14
```

Figure 9-24. *Predicted values using the linear model*

The created model can be saved in a local path and loaded again to perform the
predictions (see Figure 9-25).

```
# save fitted model to input path
path <- "C:/Users/Administrator/Desktop/linearModel"
write.ml(linearModel, path)

# read back the saved model and print
savedModel <- read.ml(path)
```

```
> # save fitted model to input path
> path <- "C:/Users/Administrator/Desktop/linearModelPath"
> write.ml(linearModel, path)
> #  read back the saved model and print
> savedModel <- read.ml(path)
> summary(savedModel)

Saved-loaded model does not support output 'Deviance Residuals'.

Coefficients:
            Estimate   Std. Error      t value   Pr(>|t|)
(Intercept)        1   2.6534e-15   3.7687e+14          0
x                  1   5.7902e-16   1.7271e+15          0

(Dispersion parameter for gaussian family taken to be 6.705318e-30)

    Null deviance: 2.0000e+01   on 3   degrees of freedom
Residual deviance: 1.3411e-29   on 2   degrees of freedom
AIC: -254.1
```

Figure 9-25. *Predicted values using the linear model*

Logistic Regression

The logistic regression is used to predict the categorical response. The `spark.ml` logistic regression can be used to predict a binary outcome (either 0 or 1) by using binomial logistic regression.

`spark.logit (data, formula, ...)` fits the logistic regression model against a SparkDataFrame. It supports `"binomial"`. Also, the model can be printed, predictions can be done on the produced model, and it can be saved to the input path.

The following example shows how to train binomial logistic regression models for binary classification to predict the categorical response. Create the data set shown in Figure 9-26 in a file named `matchDetails.txt`.

```
outlook,temp,humidity,played
sunny,hot,high,0
sunny,hot,high,0
overcast,hot,high,1
rainy,mild,high,1
rainy,cool,normal,1
rainy,cool,normal,0
overcast,cool,normal,1
sunny,mild,high,0
sunny,cool,normal,1
rainy,mild,normal,1
sunny,mild,normal,1
overcast,mild,high,1
overcast,hot,normal,1
rainy,mild,high,0
```

Figure 9-26. `matchDetails.txt` file

The data set contains four variables—`outlook`, `temp`, `humidity`, and `play`—that explained whether the match is played or not based on outlook, temperature, and humidity conditions. `Play` is the response variable and the other three columns are independent variables.

Now, build a logistic regression model to predict whether the match would be played or not based on the independent variables. Read the data from the file using the read.csv method and create a SparkDataFrame (see Figures 9-27 and 9-28).

```
if (nchar(Sys.getenv("SPARK_HOME")) < 1) {
  Sys.setenv(SPARK_HOME = "C:/Users/Administrator/Desktop/spark-2.3.0")
}

library(SparkR, lib.loc = c(file.path(Sys.getenv("SPARK_HOME"), "R", "lib")))

sparkR.session(master = "local[*]",
               sparkHome =Sys.getenv("SPARK_HOME"),
               enableHiveSupport = FALSE,
               sparkConfig = list(spark.driver.memory = "2g")
               )

filePath = "C:/Users/Administrator/Desktop/matchDetails.txt"

dataFrame = read.csv(filePath,header = TRUE)

trainingData <- createDataFrame(dataFrame)
```

```
> dataFrame
     outlook temp humidity played
1      sunny  hot     high      0
2      sunny  hot     high      0
3   overcast  hot     high      1
4      rainy mild     high      1
5      rainy cool   normal      1
6      rainy cool   normal      0
7   overcast cool   normal      1
8      sunny mild     high      0
9      sunny cool   normal      1
10     rainy mild   normal      1
11     sunny mild   normal      1
12  overcast mild     high      1
13  overcast  hot   normal      1
14     rainy mild     high      0
```

Figure 9-27. *Input data to build the logistic regression model*

```
> trainingData
SparkDataFrame[outlook:string, temp:string, humidity:string, played:int]
```

Figure 9-28. *SparkDataFrame*

Now, build the logistic model using `spark.logit`.

```
logisticModel <- spark.logit(trainingData, played ~ .)
```

```
summary <- summary(logisticModel)
```

```
print(summary)
```

Note   `played ~ .` refers to all the columns (i.e., `played ~ outlook + temp + humidity`).

Figure 9-29 shows the coefficients of the logistic regression model.

```
# fitted values on training data
fittedModel <- predict(logisticModel, trainingData)
```

```
head(select(fittedModel,"outlook","temp","humidity", "prediction"))
```

```
> print(summary)
$coefficients
                  Estimate
(Intercept)      -34.81937
outlook_rainy     47.57223
outlook_sunny     47.57223
temp_mild        -28.27543
temp_cool        -13.44601
humidity_high     16.21571
```

Figure 9-29. *Coefficients of the logistic regression model*

Figure 9-30 shows the prediction.

```
> head(select(fittedModel,"outlook","temp","humidity", "prediction"))
    outlook temp humidity prediction
1     sunny  hot     high          0
2     sunny  hot     high          0
3  overcast  hot     high          1
4     rainy mild     high          0
5     rainy cool   normal          1
6     rainy cool   normal          1
```

Figure 9-30. *Prediction using the logistic regression model*

Decision Tree

spark.decisionTree fits a decision tree regression model or a classification model on the SparkDataFrame.

Use spark.decisionTree {SparkR}. Read the same data that were created in the previous example from the file matchDetails.txt using the read.csv method and create a SparkDataFrame as shown here.

```
if (nchar(Sys.getenv("SPARK_HOME")) < 1) {
  Sys.setenv(SPARK_HOME = "C:/Users/Administrator/Desktop/spark-2.3.0")
}

library(SparkR, lib.loc = c(file.path(Sys.getenv("SPARK_HOME"), "R", "lib")))

sparkR.session(master = "local[*]",
               sparkHome =Sys.getenv("SPARK_HOME"),
               enableHiveSupport = FALSE,
               sparkConfig = list(spark.driver.memory = "2g")
)

filePath = "C:/Users/d.a.ganesan/Desktop/matchDetails.txt"

dataFrame = read.csv(filePath,header = TRUE)

trainingData <- createDataFrame(dataFrame)
```

Now create the decision tree model using `spark.decisionTree`.

```
# Fit a DecisionTree classification model with spark.decisionTree
decisionTreemodel <- spark.decisionTree
                     (trainingData, played ~ . , "classification")

summary <- summary(decisionTreemodel)

print(summary)
```

Refer to Figure 9-31 for the decision tree summary.

```
> print(summary)
Formula:  played ~ .
Number of features: 5
Features:  outlook_rainy outlook_sunny temp_mild temp_cool humidity_high
Feature importances:  (5,[0,1,2,4],[0.19892473118279586,0.4354838709677421,0.07526881720430112,0.29032258064516103])
Max Depth:  5
 DecisionTreeClassificationModel (uid=dtc_fb1cc15eb071) of depth 3 with 11 nodes
  If (feature 4 in {0.0})
   If (feature 0 in {0.0})
    Predict: 0.0
   Else (feature 0 not in {0.0})
    If (feature 2 in {1.0})
     Predict: 0.0
    Else (feature 2 not in {1.0})
     Predict: 0.0
  Else (feature 4 not in {0.0})
   If (feature 1 in {0.0})
    If (feature 0 in {0.0})
     Predict: 0.0
    Else (feature 0 not in {0.0})
     Predict: 0.0
   Else (feature 1 not in {0.0})
    Predict: 1.0
```

Figure 9-31. *Summary of decision tree model*

The prediction can be done using the `predict` method as shown in Figure 9-32.

```
predictions <- predict(decisionTreemodel, trainingData)
head(predictions)
```

```
> head(predictions)
   outlook temp humidity played                     rawPrediction                      probability prediction
1    sunny  hot     high      0 <environment: 0x000000001cb9c818> <environment: 0x000000001cbc2368>          0
2    sunny  hot     high      0 <environment: 0x000000001cba3258> <environment: 0x000000001cbc7e48>          0
3 overcast  hot     high      1 <environment: 0x000000001cba8d70> <environment: 0x000000001cbcf078>          1
4    rainy mild     high      1 <environment: 0x000000001cbae0d0> <environment: 0x000000001cbd4b90>          1
5    rainy cool   normal      1 <environment: 0x000000001cbb4b48> <environment: 0x000000001cbd9ef0>          1
6    rainy cool   normal      0 <environment: 0x000000001cbba660> <environment: 0x000000001cbe0968>          1
```

Figure 9-32. *Prediction of decision tree*

Points to Remember

- SparkR is an R package that allows us to use Apache Spark from R.

- Spark provides a distributed DataFrame, which is like R data frames to perform select, filter, and aggregate operations on large data sets.

- SparkR also supports distributed machine learning algorithm using MLlib.

In the next chapter, we discuss real-time use cases for Spark.

CHAPTER 10

Spark Real-Time Use Case

In the previous chapters, the fundamental components of Spark such as Spark Core, Spark SQL, Spark Streaming, Structured Streaming, and Spark MLlib have been covered. In this chapter, we discuss one simple real-time use case to understand how we can use Spark in real-time scenarios.

The recommended background for this chapter is an understanding of Spark fundamentals. The mandatory prerequisite for this chapter is completion of the previous chapters. Also, it is assumed that you have practiced all the demos and completed the hands-on exercises given in the previous chapters.

By end of this chapter, you will be able to do the following:

- Understand the industry applications of Spark.

- Understand the data analytics project architecture.

- Understand the real-time use cases and the need for Spark Streaming.

Note It is recommended that you read the complete chapter and understand the scenarios, where Spark is used in real time.

© Subhashini Chellappan, Dharanitharan Ganesan 2018
S. Chellappan and D. Ganesan, *Practical Apache Spark*, https://doi.org/10.1007/978-1-4842-3652-9_10

Data Analytics Project Architecture

Let's examine the project architecture, shown in Figure 10-1.

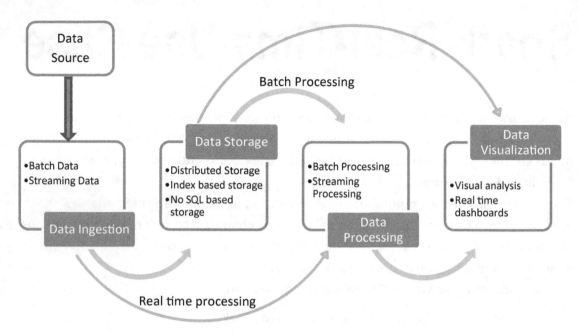

Figure 10-1. *Project architecture stages*

Data Ingestion

Data ingestion is the first layer of the project architecture, where the data are collected from multiple sources and stored in the storage layer or processed immediately. In simple words, it is defined as the process of bringing the data to the storage and processing system. The data can be ingested in batches or streamed in real time. Data ingestion parameters include the velocity of data, size and frequency of data arrival, and the format of data such as structured, semistructured, or unstructured data. The data are ingested as chunks of data at a regular time interval in the batch ingestion. The effective data ingestion is obtained by prioritizing the data sources and routing the data to the correct data storage (i.e., destination), as shown in Figure 10-2.

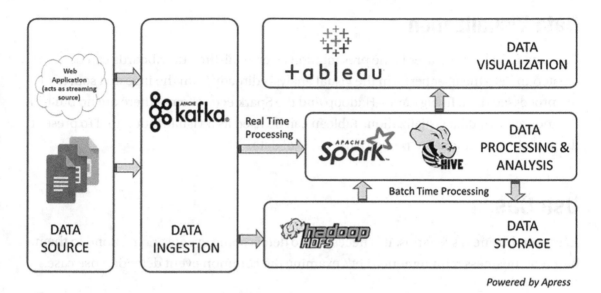

Powered by Apress

Figure 10-2. *Project architecture components in various stages*

Data Storage

Data storage becomes challenging when the volume of data increases. Data storage layers focus on how to store a huge volume of data effectively, which provides faster read and write operations for processing engines.

The storage layer should take care of storing any type of data and keep scaling to keep up with the growth of data exponentially. It should provide higher input and output operations per second (IOPS) to provide faster data delivery to the processing layer components.

Data Processing

Active analytic processing happens in the data processing layer. Data processing can be done as batch processing and real-time processing. A batch processing system gets data from the storage layer, where the batches of data were ingested, and it is applicable for offline analytics. A real-time processing system connects directly to the data ingestion layer and it is applicable for online analytics; it should provide low-latency processing results.

Data Visualization

The data visualization layer is the presentation layer. Real-time dashboards can be created to help the user perform the visual analysis directly from the ingested source or the processed data. In big data—Hadoop and the Spark ecosystem—there are no built-in components for data visualization. Tableau can be used as a visualization tool to present the real-time dashboard to perceive the value of data.

Use Cases

In certain business scenarios, it is necessary to detect some events and respond to them based on business requirements. Let's examine the common event detection use case.

Event Detection Use Case

Spark Streaming helps to detect and quickly respond to any unusual behaviors or changes in the input data pattern. For example, financial applications use triggers to detect fradulent transactions and stop fraud in a real-time manner.

The event detection use case that follows was designed for Apress publications to trigger an event if ther are any user mentions in the Twitter feed for the official Apress account, @apress. Figure 10-3 displays the architecture for implementation.

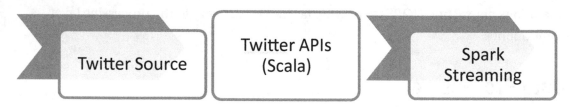

Figure 10-3. *Event detection: Twitter source*

The application is designed to fetch the tweets from Twitter based on a specified hashtag or user account by using the Twitter APIs for Spark Streaming. The Spark Streaming application processes the tweets and triggers any events if there are any described user mentions in the tweets.

The Twitter APIs for Spark Streaming are available in the following Java archives (jars).

1. `spark-streaming-twitter_2.11-1.6.3.jar`

2. `twitter4j-core-4.0.6.jar`

3. `twitter4j-stream-4.0.6.jar`

Note   The versions of jar files can be changed as needed.

The application needs some authentication tokens and secret keys to connect the Twitter source: the Consumer key, Consumer secret, Access token, and Access token secret. These keys can be obtained by creating an application in `https://apps.twitter.com/` by logging in using Twitter credentials, as shown in Figure 10-4.

Figure 10-4. *Twitter apps login*

Next, create a new Twitter app, as shown in Figure 10-5.

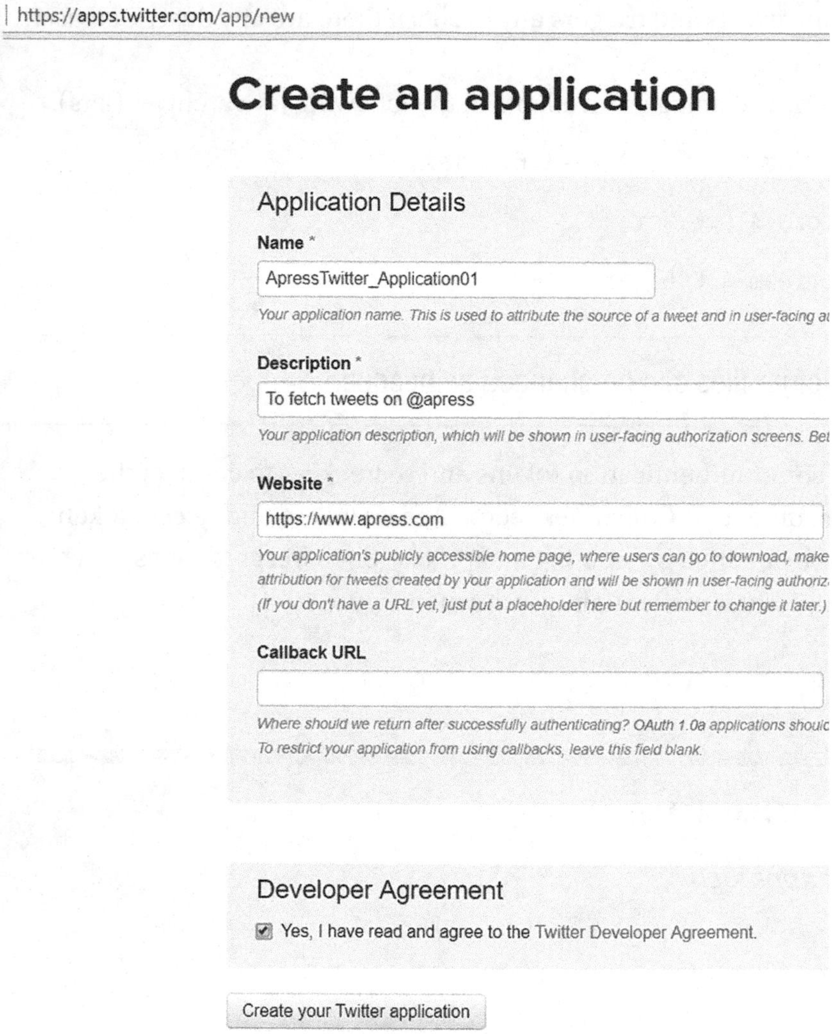

Figure 10-5. *Twitter app creation*

Once the application is created, the keys can be obtained in the Keys and Access Tokens section shown in Figure 10-6.

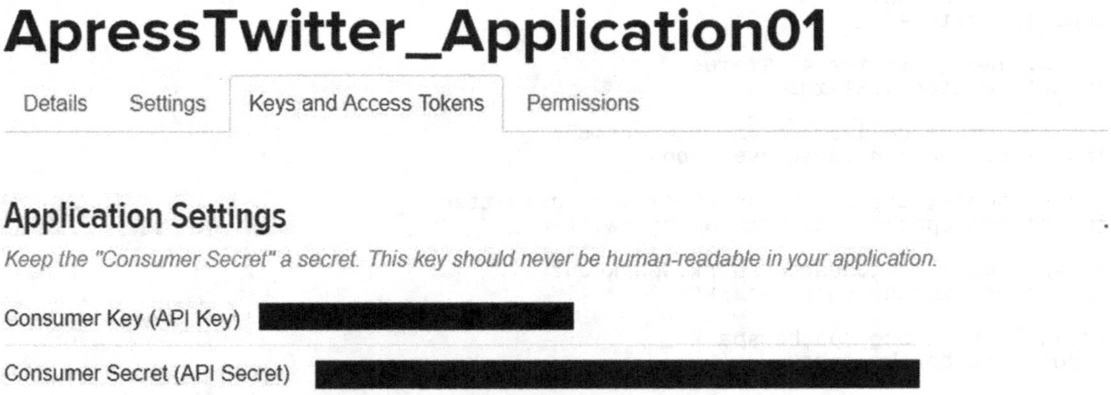

ApressTwitter_Application01

Details   Settings   Keys and Access Tokens   Permissions

Application Settings

Keep the "Consumer Secret" a secret. This key should never be human-readable in your application.

Consumer Key (API Key)   █████████████████████████

Consumer Secret (API Secret)   ████████████████████████████

Figure 10-6. *Keys and Access Tokens tab*

Figure 10-7 shows the access tokens created to authenticate the application.

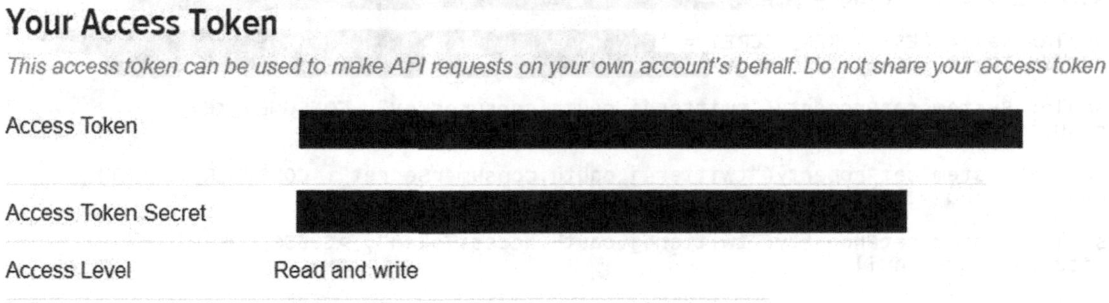

Your Access Token

This access token can be used to make API requests on your own account's behalf. Do not share your access token

Access Token   ████████████████████████████████████

Access Token Secret   ██████████████████████████████████

Access Level   Read and write

Figure 10-7. *Twitter apps access keys and tokens*

Once the authentication keys are created, add the mentioned jars to the Spark classpath and import the APIs shown in Figure 10-8.

```
scala> import twitter4j._
import twitter4j._

scala> import twitter4j.Status
import twitter4j.Status

scala> import collection.JavaConversions._
import collection.JavaConversions._

scala> import org.apache.spark.streaming.twitter._
import org.apache.spark.streaming.twitter._

scala> import org.apache.spark.SparkConf
import org.apache.spark.SparkConf

scala> import org.apache.spark._
import org.apache.spark._

scala> val CONSUMER_KEY = "█████████████████████"
CONSUMER_KEY: String = █████████████████████

scala> val CONSUMER_SECRET = "█████████████████████████████████████"
CONSUMER_SECRET: String = █████████████████████████

scala> val ACCESS_TOKEN = "███████████████████████████████████"
ACCESS_TOKEN: String = ████████████████████████████████

scala> val ACCESS_TOKEN_SECRET = "█████████████████████████████████"
ACCESS_TOKEN_SECRET: String = ████████████████████████████████

scala> System.setProperty("twitter4j.oauth.consumerKey", CONSUMER_KEY)
res0: String = null

scala> System.setProperty("twitter4j.oauth.consumerSecret", CONSUMER_SECRET)
res1: String = null

scala> System.setProperty("twitter4j.oauth.accessToken", ACCESS_TOKEN)
res2: String = null

scala> System.setProperty("twitter4j.oauth.accessTokenSecret", ACCESS_TOKEN_SECRET)
res3: String = null
```

Figure 10-8.   *Importing the APIs*

After setting all the properties and the authentication keys, create the Twitter instance and search the tweets for @apress as shown in Figure 10-9.

```
scala> val twitterInstance = new TwitterFactory().getInstance

scala> val tweets = twitterInstance.search(new Query("@apress")).getTweets

scala> tweets.foreach(tweet => println(tweet.getText + "\n"))
```

Figure 10-9.   *Searching the tweets for @apress*

The getText function (see Figure 10-10) retrieves the tweet text from all the tweets.

```
Today's @Apress $9.99 eBook (Apr. 20): "Pro Oracle Identity and Access Managemen
t Suite" by Kenneth Ramey? https://t.co/Zl78velbqi

We're thinking about #OAweek which is just 6 months away! Check out our blog fro
m last time about #openaccess where? https://t.co/C7snOhiGGD

RT @Apress: Interested in Machine Learning? Register for our first Apress Webina
r where author & #ML scientist @geoffHulten  discusses what?

RT @SN_OAbooks: We're halfway to the next #OAweek - just 6 months away! Did you
catch our blog about #openaccess last time? Editors from @S?

RT @LouiseEditor: "The Four Horsemen of the Test Suite: Stubs, mocks, spies and
dummies" - New blog post from @Apress author @deleteman123?

RT @LouiseEditor: Exciting times at @Apress as we are accepting video proposals!
 Contact me for more details: louisecorrigan@apress.com -us?

RT @SN_OAbooks: We're halfway to the next #OAweek - just 6 months away! Did you
catch our blog about #openaccess last time? Editors from @S?
```

Figure 10-10. *The getText function results*

The complete code for this use case is given next and the build procedure is also explained.

```scala
package com.apress.twitteranalysis

import twitter4j._
import twitter4j.Status
import collection.JavaConversions._
import org.apache.spark.streaming.twitter._
import org.apache.spark.SparkConf
import org.apache.spark._

object ApressTwitterTweetsEventDetection {

  def main(args: Array[String]): Unit = {

    val sparkSession = SparkSession.builder
                    .appName("ApressEventDetectionExample")
                    .master("local[*]")
                    .getOrCreate()
```

```
import sparkSession.implicits._

val CONSUMER_KEY = "<Specify your key>"
val CONSUMER_SECRET = "<Specify your key>"
val ACCESS_TOKEN = "<Specify your key>"
val ACCESS_TOKEN_SECRET "<Specify your key>"

System.setProperty("twitter4j.oauth.consumerKey", CONSUMER_KEY)
System.setProperty("twitter4j.oauth.consumerSecret", CONSUMER_SECRET)
System.setProperty("twitter4j.oauth.accessToken", ACCESS_TOKEN)
System.setProperty("twitter4j.oauth.accessTokenSecret", ACCESS_
TOKEN_SECRET)

val twitterInstance = new TwitterFactory().getInstance

val tweets = twitterInstance.search(new Query("@apress")).getTweets

tweets.foreach(tweet => println(tweet.getText + "\n"))

  }
}
```

Note   To execute the given code in any IDE that supports Scala, it is mandatory to add the Scala library to the project workspace and all the Spark jars to the classpath.

In this example, we just print the tweets in the console. Now it can be checked for any word or any event in the tweet and any custom implementation can be triggered based on the content of the tweets.

Build Procedure

The jars files can be downloaded from the Maven central repository or using SBT to build the application.

spark-streaming-twitter_2.11-1.6.3.jar

```
http://central.maven.org/maven2/org/apache/spark/spark-streaming-
twitter_2.11/1.6.3/spark-streaming-twitter_2.11-1.6.3.jar
```

The dependency is added in SBT as follows:

```
libraryDependencies += "org.apache.spark" %% "spark-streaming-twitter"
%"1.6.3"
```

twitter4j-core-4.0.6.jar

```
http://central.maven.org/maven2/org/twitter4j/twitter4j-core/4.0.6/
twitter4j-core-4.0.6.jar
```

The dependency is added in SBT as follows:

```
libraryDependencies += "org.twitter4j" % "twitter4j-core" % "4.0.6"
```

twitter4j-stream-4.0.6.jar

```
http://central.maven.org/maven2/org/twitter4j/twitter4j-stream/4.0.6/
twitter4j-stream-4.0.6.jar
```

The dependency is added in SBT as follows:

```
libraryDependencies += "org.twitter4j" % "twitter4j-stream" % "4.0.6"
```

Building the Application with SBT

The SBT installation procedure has already been discussed in the previous chapters. Follow the further steps to add the Twitter and Spark Streaming dependencies in the build.sbt file as shown here. Add the content shown in Figure 10-11 in the build.sbt file.

```
name := "ApressTwitterTweetsEventDetection"

version := "1.0"

scalaVersion := "2.11.8"

libraryDependencies += "org.apache.spark" %% "spark-streaming-twitter" % "1.6.3"

libraryDependencies += "org.twitter4j" % "twitter4j-core" % "4.0.6"

libraryDependencies += "org.twitter4j" % "twitter4j-stream" % "4.0.6"
```

Figure 10-11. *build.sbt file*

SBT downloads the required dependencies for the Spark SQL and keeps them in the local repository if it is not available while building the jar.

Note   It is recommended to use the SBT for building and packaging the Scala classes.

Create the folder structure as displayed in Figure 10-12 for the SBT build.

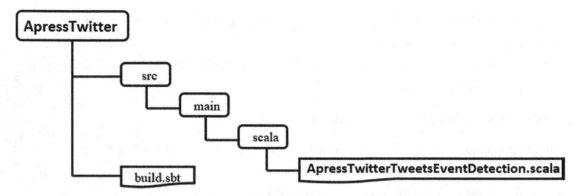

Figure 10-12. *Directory structure for SBT build*

ApressTwitter is the parent directory and src/main/scala are subdirectories. Navigate to the folder ApressTwitter (i.e., cd /home/ApressTwitter). Now execute the Scala build package command to build the jar file.

```
> cd /home/ApressTwitter
> sbt clean package
```

Once the build has succeeded, it creates the project and target directory as shown in Figure 10-13.

Name		Type	Size
project		File folder	
src		File folder	
target		File folder	
build.sbt		SBT File	1 KB

Figure 10-13. *SBT build directory structure*

SBT creates the application `ApressTwitterTweetsEventDetection-1.0_2.11.jar` in the target directory. Now, the application can be submitted to the Spark cluster by using this command:

```
spark-submit --class ApressTwitterTweetsEventDetection --master
spark://<hostIP>:<port> ApressTwitterTweetsEventDetection-1.0_2.11.jar
```

where `spark://<hostIP>:<port>` is the URI for Spark master. By default, the Spark master runs on port 7077. However, it can be changed in the configuration files.

Points to Remember

- Data analytics life cycle and layers are data ingestion, data storage, data processing, and data visualization.

- Data ingestion is the process of extracting data from multiple sources (batch or real-time) and persisting in the storage layer.

- The storage layer should take care of storing any type of data and keep scaling to keep up with the growth of data exponentially.

- The data processing system connects directly to the data ingestion layer or storage layer and it is applicable for online and offline analytics to provide faster and low-latency processing results.

- Dashboards can be created to help the user to perform visual analysis directly from the ingested source or the processed data in the visualization layer.

Index

A

Apache Spark, 60, 141, 237
    installation, 73–76
    master UI, 76–77
    persisting RDD, 104
    prerequisites, 71–73
    scala code, 109
    storage levels, 104
Apache Zookeeper, 178
Application programming
       interfaces (APIs), 80

B

Batch processing, 142, 263

C

Currying function, 31

D

dapplyCollect function, 249
dapply function, 248
Data analytics project architecture
    components, 263
    data ingestion, 262
    processing data, 263
    stages, 262
    storage, 263
    visualization, 264

DataFrames
    creation, 117
        JSON content, 118
        show() method, 118
    operations, 118
        filter() transformation, 119
        groupBy() transformation, 120
        select() transformation, 119
    view, creation, 121–122
Data ingestion, 262
Data processing, 142, 263
Datasets
    BookDetails.json, 123
    operations, 123–124
    reflection-based approach, 125
        class attributes, 127
        DataFrame, creation, 126
        RDD, creation, 125
        schema creation, 128
Data storage, 263
Data streaming, 142
Decision tree regression model
    creation, 259
    predict method, 259
    SparkDataFrame, 258
    spark.decisionTree {SparkR}, 258
Direct Acylic Graph (DAG), 79
    lineage graph, 104
    scheduler, 101
    visualization, 103
Discretized Streams (DStream), 144

© Subhashini Chellappan, Dharanitharan Ganesan 2018
S. Chellappan and D. Ganesan, *Practical Apache Spark*, https://doi.org/10.1007/978-1-4842-3652-9

E

Event detection
    Apress publications, 264
    code, 269–270
    getText function, 269
    import API, 268
    Spark streaming, 264
    tweets, @apress, 268
    Twitter APIs, 265
    Twitter apps, 265
        creation, 266
        keys and tokens, 267
    Twitter source, 264–265

F, G

Fault tolerance, 87
Full-session-based tracking, 152–154
Functional programming (FP), 2
    anonymous function, 27–28
    function composition, 30
    function currying, 31
    higher order functions, 29
    nested functions, 32–34
    pure function, 2
        example, 3
        and impure function, 4
    variable length
        parameters, 34–36
Function currying, 31–32
Fundamental components,
    Spark, 261

H

Hadoop Distributed File System
    (HDFS), 80, 85, 97, 143
Hive metastore, 243

I, J

Immutability, 87
In-memory computation, 87
Input and output operations
    per second (IOPS), 263
Integer indices, 212

K

Kafka
    APIs, 176–177
    architecture, 178
    cluster, 180
    concepts, 177–178
    consumer console, 182
    distributed streaming platform, 175
    folder, 181
    integration
        Spark application, 182
        Spark structured streaming, 185
    partitioned log, 179

L

Linear regression, 224
    fitted model, 251
    predict function, 253
    predict values, 254
    R environment, 251
    Spark DataFrame, 252
    spark.glm {SparkR} package, 250
    Spark libraries, 252
    write.ml(model,path), 251
Logistic regression, 229
    binomial, 255
    coefficients, 257
    prediction, 258
    read.csv method, 256

SparkDataFrame, 256–257
spark.logit, 255
spark.ml, 255

M, N, O

Machine algorithms, 250
Machine learning (ML), 190
Maven central repository, 270
ML pipelines
    classification algorithms, 229
    creation, 219
    estimator, 216
    importing, APIs, 217
    K-Means clustering algorithm, 234–235
    predicting tables, 220, 222
    Spark Shell, 217
    test documents, 219
    testing time usage, 223
    training time usage, 223
    transformer, 216
Multinode cluster setup
    Oracle VirtualBox (*see* VirtualBox)
    Spark, 60
        application UI, 68
        installation, 62
        master UI, 67
        prerequisites, 61
        stopping, Spark cluster, 70

P, Q

Partitioning, 87
Pattern matching, 13–14
Pearson chi-square (χ^2) tests, 198–199
Pearson correlation, 195–196
Product–moment correlation coefficient
    (PMCC), 195

Pure function, 2
    example, 3
    and impure function, 4

R

read.df method, 242
Read, evaluate, print loop (REPL), 85
Real-time processing, 80, 142
Regression algorithms, 224
Relational database management system
    (RDBMS), 121
Resilient distributed data set (RDD)
    actions, 87
        count(), 96
        first(), 96
        foreach(func), 97
        foreach(println), 98
        reduce(func), 95
        result data set, 96
        take(n), 97
    clusters, 86
    creation
        Hadoop File System, 90
        parallelize method, 88
        partitioning, 90
        textFile() method, 89
    fault tolerance, 87
    immutability, 87
    in-memory computation, 87
    lazy evaluations, 87
    operations, 88
    partitioning, 87
    transformations, 87
        distinct([numTasks]), 94
        filter(func), 92
        flatMap(func), 91
        intersection(otherDataset), 94

Resilient distributed data set (RDD) (*cont.*)
        map(func), 91
        mapPartitions(func), 92–93
        mapPartitionsWithIndex(func), 93
        union(otherDataset), 93–94
    variables, 105
        accumulators, 106–107
        broadcast, 106
RStudio, 238

S

Scala programming
    case classes, 18, 20–21
    class *vs.* object, 14–15
    companion classes and
        objects, 17–18
    getOrElse() method, 23
    immutablity, 7
    iterating over collection, 23–24
    lazy evaluation, 8, 10
    methods of collection, 25–26
    Option[T] collection, 23
    pattern matching, 13–14
    singleton object, 15–16
    string interpolation (*see* String
        interpolation)
    type inference, 6
    variable declaration and
        initialization, 5
Simple Build Tool (SBT), 79, 270–271
    ApressTwitterTweetsEvent
        Detection-1.0_2.11.jar, 273
    build directory structure, 272
    build.sbt file, 271
    Spark cluster, 273
    Spark master, 273
    Spark SQL, 272

Spark
    architecture, 82
    cluster manager, 83
    components, 84
        data frame distribution, 85
        GraphX, 85
        MLib, 85
        RDD, 84
        SQL data, 84
        streaming, 85
    DAG (*see* Direct Acylic
        Graph (DAG))
    vs. Hadoop MapReduce, 81
    pair RDDs
        groupByKey([numTasks]), 98
        reduceByKey (func,
            [numTasks]), 99
        sortByKey([ascending],
            [numTasks]), 100–101
    SBT, 107
        folder structure, 108
        output directory, 111
        Spark cluster, 110
        target directory, 109
Spark binaries, 62
SparkDataFrame
    column operations, 247
    creation
        .csv file, 242–243
        data sources, 242
        Hive table, 243
        JSON file, 242
        local R DataFrame, 241
        read.df method, 242
    defined, 237
    grouping and
        aggregation, 245–246
    select query, 244–245

Spark environment
    variable, 240
Spark Machine Learning Library
    (Spark MLlib), 189
    correlation, 195
    DataFrame-based APIs, 190
    features, 190
        extraction, 201
        StopWordsRemover, 207–209
        StringIndexer, 209–211
        tokenizer, 206
        VectorSlicer, 212–215
    hypothesis testing, 198
    pipelines (*see* ML pipelines)
    sparse vectors, 193–194
    TF–IDF, 201, 203, 205
    vectors in Scala, 191–192
Spark master, 65, 67, 273
SparkR
    dapplyCollect
        function, 249
    dapply function, 248
    RStudio, 238–240
    sparkR.session, 240
    SQL queries, 249–250
Spark single-node cluster
    setup, 70–71
SparkSession, 116, 238
    Hive support, 133
Spark Shell, 85–86, 117, 185, 217
Spark SQL
    DataFrame (*see* DataFrames)
    dataset (*see* Datasets)
    data sources, 129
        format() functions, 130
        JDBC connectivity, 132
        load/save functions, 129
    Hive tables, 133–135

SBT building, 135
    cluster, 137
    directory structure, 137
    folder structure, 136
Spark streaming, 142, 264
    architecture, 143
    DStream, 144
    features, 143
    internal working, 143
    stateful streaming, 149
        applications, 155
        full-session-based tracking, 152–154
        window-based
            streaming, 149–152
    Streaming Context, 144
    using TCP socket, 145–148
Spark structured streaming
    DataFrames/Datasets (*see* Streaming
        DataFrames/Datasets)
    definition, 158
    programming model, 158–159
        word count example, 160–162
    stateful streaming
        watermarking, 170–171
        window operations, 167–170
    triggers, 171
        fault tolerance, 173
        type, 172–173
Sparse vectors, 193–194
Spearman correlation, 195, 198
Streaming DataFrames/Datasets
    creation, 163–164
    operation, 164–167
String indices, 212
String interpolation, 10
    f interpolator, 12–13
    raw interpolator, 13
    s interpolator, 11

T, U

Tableau tool, 264
TCP socket, 143, 145–148, 160
Term Frequency–Inverse Document
  Frequency (TF–IDF), 201, 203, 205

V, W, X, Y, Z

VirtualBox
  installation, 41–47
  manager, 40
  SparkMaster machine, 59–60

virtual machine creation, 49
  hard disk, creation, 51–52
  hard disk file location,
    specification, 54
  hard disk storage type, selection, 53
  iso disk file, 56
  memory specification, 50
  network adapter selection, 57
  network configuration, 58
  network settings, 57
  settings, 56
welcome page, 48

Printed in the United States
By Bookmasters